Packaging for 3D Assembly of Diverse Systems

Claire

Packaging for 3D Assembly of Diverse Systems

First Edition November 2024

Copyright © 2024 Claire

Written by Claire

CONTENTS

Chapter 1 1
Introduction 1
   1.1. Three-Dimensional Integration 2
      1.1.1. 3D Integration: Manufacturing Methods 2
   1.2. 3D IC Technology Landscape 3
      1.2.1. Package-level 3D integration 4
      1.2.2. Chip-level 3D integration 5
      1.2.3. Within-die 3D integration 6
      1.2.4. Monolithic 3D Integration 7
   1.2. 3D Heterogenous Integration 8
   1.3. Thesis Motivation and Brief Outline 9
   1.4. Thesis Organization 10

Chapter 2 13
Stacking of the Ultra-Thin Silicon layers with functional MOS Devices and its reliability . 13
   Introduction 14
   2.1. Fabrication 15
      2.1.1. Transistor fabrication NMOS and PMOS 15
      2.1.2. Ultra-thin silicon transfer using epoxy and Au-In TLP bonding 15
   2.2 Vertical Stacking 19
      2.2.1. Process flow 19
      2.2.2. Keep out zone 21
   2.3 Characterization of the transferred devices 22
      2.3.1. Diffusion measurements 22
      2.3.2. DC electrical measurements 22
      2.3.3. Negative differential resistance 24
      2.3.4. 3-layer stack with functional devices and DC measurements 28
      2.3.5. 5-layer Ultra-thin silicon stack 29

2.4 Reliability measurements on the Ultra-Thin Silicon Stack 29

    2.4.1. Fabrication of two-layer stacks with functional devices 29

    2.4.2. Reliability tests 29

    2.4.3. DC electrical measurements 30

Conclusion 32

Chapter 3 33

Heterogenous Stacking of III-nitride-on-Si HEMTs and Si-MOSFETs to Copper 33

Introduction 34

3.1. Transfer to glass using epoxy bonding method 35

    3.1.1. Growth and transistor fabrication AlGaN/GaN HEMT 35

    3.1.2. Temporary bonding and silicon removal 36

    3.1.3. Permanent bonding using epoxy 37

    3.1.4. DC electrical measurements 37

    3.1.5. JEDEC reliability tests 38

3.2. Transfer to glass using TLP Au-In bonding method 40

    3.2.1. Permanent bonding using Au-In bond 40

    3.2.2. DC electrical measurements 40

3.3. Transfer to copper using TLP Cu-In bonding method 41

    3.3.1. Growth and transistor fabrication HEMT and NMOS 41

    3.3.2. Temporary bonding and silicon removal 41

    3.3.3. Permanent bonding using Cu-In bond to Copper 42

    3.3.4. DC electrical measurements 43

Conclusion 46

Chapter 4 48

3D Integration of Heterogeneous Dies for Fluorescent Detection 48

Introduction 49

4.1. Individual components fabrication 50

    4.1.1. Photodetector Fabrication 52

    4.1.2. Selection of optical filter 53

    4.1.3. Fabrication of the glass fluidic chip with microheater 55

4.2. Hybrid Integration 55

    4.2.1. Device Stacking by epoxy bonding 55

    4.2.2. Bonding plastic filters to silicon photodetector 56

    4.2.3. Planarization of the bonded filter stack 57

    4.2.4. Bonding of the glass fluidic chip 59

    4.2.5. Planarization of the fluidic chip and polymer via opening 60

  4.2.6. Interconnecting the components in the stack 61
 4.3. Device component testing 62
  4.3.1. Effect of stacking on photodetector sensitivity 62
  4.3.2. Effect of microheater proximity on photodetector sensitivity 63
  4.3.3. Platform testing 64
 Conclusion 69

Chapter 5 70

Interconnects 70

 Introduction 71
 5.1. 3D stacking by wire bonding 72
  5.1.1. Fabrication of dummy MEMS dies 72
  5.1.2. Fabrication of dummy ultra-thin silicon dies 74
  5.1.3. Stacking of the MEMS and silicon devices 75
  5.1.4. Stacking results 76
  5.1.5. Temperature cycling of the stacked devices 77
 5.2. 3D stacking by Ink jet printed metal via filling 78
  5.2.1. single layer 3D stacking and interconnection 78
  5.2.3. Combination of Ink jet printed Via fill and sputtered metal contacts 82
 Conclusion 84

Chapter 6 86

Summary and Future Directions 86

 6.1 Summary 86
 6.2 Further scope 88

Bibliography 90

Chapter 1
Introduction

"Another direction of improvement (of computing power) is to make physical machines three dimensional instead of all on a surface of a chip (2-D). That can be done in stages instead of all at once; you can have several layers and then many more layers as the time goes on." - an excerpt from talk delivered by Richard Feynman titled "Computing Machines in the Future" at the Gakushuin University in Tokyo on August 9,1985. Three decades later, advances in nanofabrication have enabled us to imagine, design, create and stack microchips for 3D Integration.

Currently, heterogeneous 3D Integration is faster than Moore (Figure. 1.1). This thesis details the development of processing technologies for the 3D packaging for integration of heterogeneous systems.

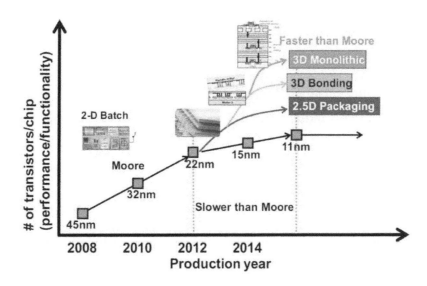

Figure 1. 1: Semiconductor and package technology road map[1]

1.1. Three-Dimensional Integration

Due to the limitations of silicon material, further scaling[2-5] of devices is much more challenging. Hence, to keep up with the Moore's law[6] (miniaturization law), there is a need for an immediate technology driver[7] and 3D integration technology is believed to be a potential candidate. In 3D integration technology, the processed or to be processed wafers/chips are stacked in the third dimension[8-12]. This technology not only provides reduction in board foot print, high speed and low power devices, but also allows the possibility to integrate multi-functional platforms by stacking[3,8,13-15]. The 3D integration technology enables heterogeneous integration with higher density[7] making it faster than Moore [1,16,17].

The use of TSVs (Through-silicon via[7,18-20]) in 3D integration makes 3D integration different from 3D packaging[8]. TSV technology involves six main steps –

- (i) Etching/drilling the silicon via using deep reactive-ion etching (DRIE) tool
- (ii) Dielectric layer deposition by the plasma-enhanced chemical vapor deposition (PECVD) technique
- (iii) Barrier and the seed layer depositions by the physical vapor deposition (PVD) techniques
- (iv) Filling of the vias using electroplating (generally copper is used to fill the vias)
- (v) Copper over - burden removal by the chemical mechanical polishing (CMP) technique and
- (vi) TSV copper revealing.

TSV offers unique advantages and enables higher number of interconnections, reduced latency, lower inductance, lower capacitance, and permit higher speed communications, and lower power level communication links between circuits[21].

1.1.1. 3D Integration: Manufacturing Methods

Out of the various 3D integration schemes followed, one of the more popular ideas is to categorize the process by die or wafer level stacking: wafer-to wafer (W2W), chip-to-chip (C2C), and chip-to-wafer (C2W). In the W2W stacking approach, processed wafers were thinned, aligned, bonded and further dies were singulated using dicing. In the C2C stacking approach processes wafers were diced and known good dies were stacked individually by the bonding process. In the C2W approach, processed wafers were diced and known good dies were identified. Further, these dies were assembled on a new wafer for reconstruction, followed by further stacking and singulation. W2W integration technique offers wafer level integration and hence has the highest throughput with the best alignment accuracy. However, only the identical size dies will be stacked and there is an inherent risk due to a defective die stacked to the good dies, there by destroying the whole stack. C2C stacking approach is not a cost-effective solution due to single die processing. However, it is a very well-known technique in high margin and high-performance device application areas like space. The C2W technique offers unique advantages like heterogenous stacking, integration of non-identical sized known good dies and integration of dies with dissimilar materials. Hence, this technique allows stacking of not only silicon-based semiconductor dies, but also the other semiconductor devices manufactured at different foundries. Figure 1. 2 shows the difference between W2W and C2W stacking approaches[22].

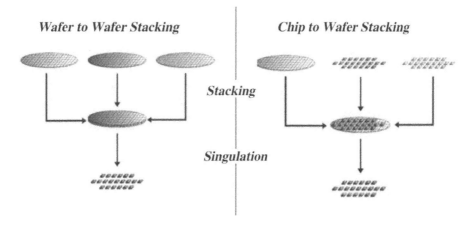

Figure 1. 2: Difference between Wafer to Wafer and Chip to Wafer stacking methodologies[22]

1.2. 3D IC Technology Landscape

As compared to 2D technologies, 3D integration technology offers many advantages like devices with better electrical performance, wider bandwidth, lower power consumption, smaller form factor, higher density, lower cost and lighter weight. Present day 3D integration technology can be classified into 4 major platforms,

(i)   Package level 3D integration

(ii)   Chip-level 3D integration or 2.5D stacked IC technology (3D SIC)

(iii) Within-die 3D integration or 3D system-on-chip (3D SOC)

(iv)   Monolithic 3D IC integration

1.2.1. Package-level 3D integration

Over the past 2 decades, many technologies for package level 3D integration have been developed and are in high-volume manufacturing[4,7,12]. Among them, the most popular technology is to stack the processed dies and interconnect them to a printed circuit board (PCB) using wire bonding technique as shown in Figure 1.3 (a) and (b). The applications include micro electro mechanical systems (MEMS) die and application specific integrated circuit (ASIC) die stacking for sensors[23] as shown in Figure 1.4(a) and the memory die stacking.

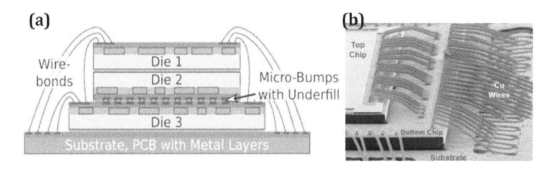

figure 1. 3: Package level 3D integration (a) Traditional die stacking with wire bonding to the PCB[24] (b) Amkor's 3d IC packaging with Copper wires[8]

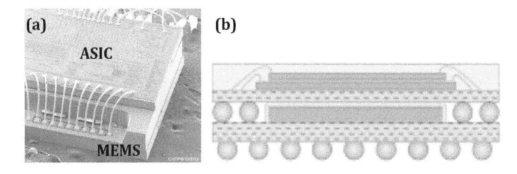

figure 1. 4: (a) MEMS and ASIC die stacking and interconnection using wire bonding[23] (b) Package on package stacking approach[7]

Chapter 1 Introduction

Another approach using package-on-package (PoP)stacking technology is shown in Figure 1.4(b). PoP is widely used in the mobile devices, where packages were stacked together to realize 3D integration. Currently, in mobile devices - memory, external LPDDR DRAM (low power double data rate dynamic random-access memory) and non-volatile flash memories are stacked[7]. Using 3D PoP, memory die and its controller are packaged and stacked independently using the ball grid array (BGA) and interconnected above a PCB[4,25] (Figure 1.5). In many devices - application processor and DRAM were using 3D PoP[25]. Another well-known approach is embedded Wafer Level Ball Grid array[26] (eWLB). This approach involves reconstituting a wafer (from the known good dies), by moulding individual die, placed in an array format into an artificial wafer[26,27].

 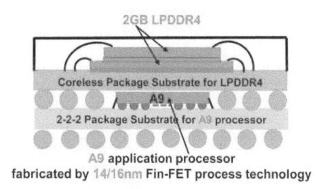

Figure 1. 5: Package-on-Package in Apple's iPhone 6s smart phone[25]

1.2.2. Chip-level 3D integration

In this approach the processed dies were stacked and interconnected; enabling high-density face-to-face connections using smaller connections (micro bumps[28]). The electrical connection to the active layer is realized through the silicon substrate of the active device. This methodology is popularly known as Through silicon via (TSV) technology[4,7,12] which is the heart of this approach. TSV's are further used for interconnecting the dies or connecting the die to a package substrate using micro bumps as shown in Figure 1.6 (a).

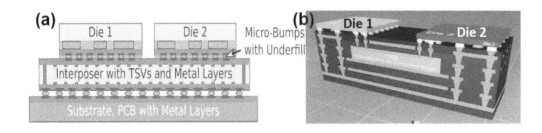

Figure 1. 6: Interposer technology (a) schematic of the interposer technology[24] (b) Intel's EMIB technology[29]

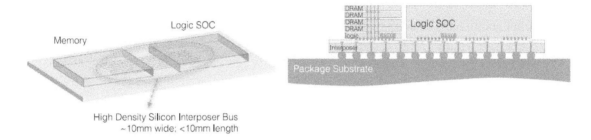

Figure 1. 7: Typical interposer technology used in 3D memory stacks with logic SOC die[7]

In order to enable high-bandwidth[30] and low latency, multiple dies[31] are stacked side-by-side[32] onto a silicon carrier wafer in this approach. The silicon carrier wafer with TSVs[33] used in this approach is called as an Interposer. The interposer wafer doesn't contain any active elements and hence this technology is popularly called as the 2.5-D technology or 3D stacked IC technology (3D SIC) (Figure 1.6 (a)). The main areas of applications include memory die stacking and interconnecting it with a logic SOC using high density interposer bus[7,25] as shown in Figure 1.7. It is also used in FPGA die stacking allowing for high speed connections. Intel's embedded multi die interconnect bridge (EMIB) technology[29] is also an example of Interposer technology wherein, silicon embedded bridge in the organic package or interposer takes care of the lateral communication between the chips as shown in Figure 1.6 (b).

1.2.3. Within-die 3D integration

Within-die 3D integration technology is also known as 3D system-on-chip (3D SOC) technology. In the VLSI system-on-chip technology, a collection of large functional circuit blocks (individual IP's) are connected at the global interconnect level. These functional blocks may have different functional requirements (digital, analog, CMOS, memory – DRAM/SRAM, etc) and need to be integrated to perform complex functions. Instead of having multiple functional blocks as individual IC's on a PCB, SoC is split into 2 or more parts and were realized with optimum 3D IC technologies at lower cost per function. Further, these functional circuit blocks were stacked and interconnected using 3D interconnect processes. An example of such technology is Amkor's Double POSSUM[25] multi stacked die configuration enabling integration of ASICs, MEMS, memory, microcontrollers, etc as shown in Figure 1.8.

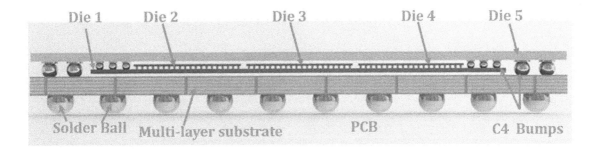

Figure 1. 8: Amkor's multi-chip-to-chip interconnects[25]

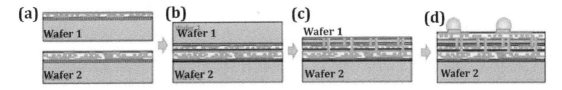

Figure 1. 9: (a) Processed device wafers (b) Room temperature alignment and bonding of both the wafers (c) Thinning of the top wafer (wafer 1) and realization of TSVs (d) Back end processing for bonding to a final PCB[7]

The fabrication methodology using this approach is illustrated in Figure 1.9. In this approach, W2W bonding methodology is used to align and bond the cleaned and plasma treated processed wafers at room temperature as shown in Figure 1.9. The top wafer can be thinned, and the silicon back-side electrical contacts were created to interconnect top and bottom wafers. In this approach, W2W overlay bonding accuracies are very critical.

1.2.4. Monolithic 3D Integration

The sequential 3D process of bonding an un-patterned semiconductor layer to processed wafer for sequential building of second layer of devices is referred to as monolithic 3D integration[7,34-42]. Even though the sequential processing of second device layer on top of processed layers technique offers higher yield, this process is severely impacted by thermal limits[7]. Hence, identical device layer stacking is not feasible using this sequential approach. This approach is appropriate for 3D heterogenous integration technologies, where dissimilar material property layers need to be stacked and interconnected using variety of interconnection technologies (not limiting to only TSV technology). An example of a monolithically integrated vertical layers can be seen in Figure 1.10.

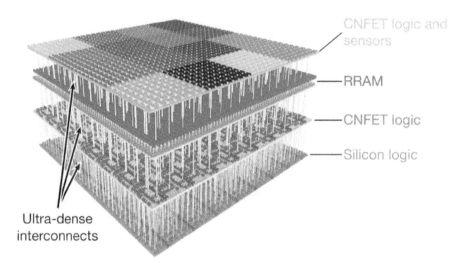

Figure 1. 10: Monolithic Integration of 4 functional layers and interconnected through dense vertical inter-connects[34]

1.2. 3D Heterogenous Integration

In the past 50 years, there has been a consistent growth in the semiconductor industry which was guided by the Moore's law[6]. Due to existing limitations in the silicon processing technologies, further scaling of devices has become technically and financially challenging[43]. Consequently, the focus has shifted towards scaling of systems for which there is a need to investigate alternative technologies. With the advent of mobile computing and communications market[44], there has been a sudden increase in demand for miniaturized systems with extended functionality and better performance. To address these needs, researchers are focusing on comparatively less explored assembly and packaging technologies[45]. As discussed in previous reports[5,10,46], over the last 45 years silicon has scaled by a factor of 1000x, whereas the printed circuit board (PCB) trace pitch, ball grid array pitch and the bump pitch to packages have scaled only 4x times. One of the approaches to address this issue is to utilize the 3rd dimension for scaling by heterogeneous integration[8-10]. Heterogeneous integration involves integration and interconnection of various components manufactured separately in a single stack, so that the integrated system can be more powerful, efficient and functionally diverse[45,49]. Currently, various technologies are being developed for heterogenous integration. Such 3D integrated heterogeneous systems would provide complex functionality[50,51], for the upcoming plethora of market demands in Internet of Things (IOT)[17,52] based smart devices with big data cloud applications[53] and Healthcare based applications[54].

1.3. Thesis Motivation and Brief Outline

Though there are several reports on 3D integration technologies, which have been for stacking and interconnection mechanisms, all these technologies are confined to silicon-based devices. With the existing processing technologies, it is difficult to integrate complex combinations of GaN-HEMT's, MEMS, microfluidics, optical devices and CMOS at the wafer-level. Moreover, due to the cross-contamination issues, most of these devices are never accepted in the standard CMOS foundries. These devices are very expensive compared to existing CMOS based devices. Also, the current through silicon via interconnection technology is not suitable for stacks having complex combinations. To address these challenges, we report development of processing technologies towards 3D Heterogenous Integration by post fab vertical stacking techniques. We demonstrate various techniques to vertically stack ultra-thin layers of different devices from different technologies fabricated using different materials.

In the first part of work, we developed fabrication techniques for the transfer of ultra-thin silicon (UTSi) layers of thickness ~1.45 μm on to a foreign substrate using low temperature bonding approach. In order to enable this transfer, we have analysed and resolved the associated stress and reported a scheme of mitigating the stress issues. This resulted in a crack-free transfer of ultra-thin silicon layer. Based on this technique, we have demonstrated three-layer stacking of the ultra-thin silicon layers with functional MOSFETs in each layer using low temperature epoxy bonding processes. Electrical characterization results of NMOS/PMOS devices in each layer has been presented and compared for before and after transfer. Due to the non-thermally conducting behaviour of the epoxy bonding, degradation in the device behaviour has been observed after the transfer process. To address this issue, metal-based Cu-In bonding process has been explored and the ultra-thin device layers were transferred to the copper substrate for better thermal conductivity. Further, we have successfully carried out stacking of heterogenous materials like GaN and silicon with functional transistors on to copper substrate. Due to the large lattice and thermal mismatch, transfer of GaN grown on silicon to foreign substrate becomes a non-trivial process. we have shown that an improvement in the performance of the GaN HEMT device, can be achieved by the new transfer process. Also, we were able to demonstrate heterogeneous stacking through post-fabrication transfer of an ultra-thin silicon layer with MOSFET on top of the GaN HEMT layer.

Further, 3D heterogenous integration of a miniaturised hybrid systems was demonstrated. In this 3D packaging technology, we were able to integrate photodetector, optical filters, microfluidic chip with micro heater and an LED. The functionality of the heterogenous system was demonstrated by measuring the increase in photodetector current due to the fluorescence property of the Rhodamine B and Rhodamine 6G. Finally, we have demonstrated an approach for interconnecting the stacked layers using the nonconventional inkjet printing technique.

1.4. Thesis Organization

The thesis work presented here focuses on developing the processing technologies that would allow 3D packaging by the post fab vertical stacking technique.

In *Chapter 2*, we present a process flow, to transfer the ultra-thin silicon device layers, of thickness as small as ~1.4 µm, to a foreign substrate, which enables vertical stacking of functional layers. The stacking is performed at a considerably low temperature of 150°C. Using the metal and the epoxy-based bonding methods we demonstrate transfer of ultra-thin silicon layers with functional devices. Further, using the epoxy bonding technique we have achieved the multi-layer stacking. We also present comparison of electrical characterization results for devices before and after the transfer.

In *Chapter 3*, we have implemented a new approach for stacking of AlGaN/GaN high electron mobility transistors (HEMTs) and Si MOSFETs onto a copper substrate. Here, both the devices are first fabricated on silicon substrate using separate process flows. Then GaN devices are transferred onto a copper substrate to improve thermal conductivity. However, large lattice and thermal mismatch makes transfer of GaN grown on silicon a non-trivial process. Thick electroplated copper is used to improve mechanical strength, allowing transfer of the GaN layer using Cu-In bonding. Next, an ultra-thin silicon layer (~1.5 µm) with functional NMOS transistors were separated from parent SOI wafer and then stacked above the GaN devices using cost-effective epoxy bonding approach. The Cu-In bonding not only improved thermal performance but also led to significantly better device behaviour.

In *Chapter 4,* we present a simple 3D integration method for miniaturisation of systems. Various components of the system were stacked using SU-8 based planarization and epoxy-based bonding. Spacer dielectric (SU-8) is patterned using photolithography for formation of interconnect vias. Electrical interconnects over the large topography between the layers is formed by screen-printing of silver nanoparticle epoxy. Using this integration technique, we realize a fluorescence sensing platform consisting of a silicon photodetector, plastic optical filters, commercial LED and a glass microheater chip. In this chapter-4, several fabrication challenges such as planarization, stacking and interconnection of these divergent chips have been tackled and resolved. For example, process incompatibility of the plastic optical filters is sorted out by additional passivation using parylene-C. The functionality of the system is demonstrated by detecting the fluorescence property of Rhodamine B and Rhodamine 6G dyes. This process flow can be scaled to stack a larger number of layers for realising more complicated systems with enhanced functionality and applications.

In *Chapter 5,* we present a methodology for interconnecting three ultra-thin (10 µm) silicon (UTSi) stacked layers. The main aim is to develop a cost-effective methodology suitable for heterogenous integration. For the UTSi layers interconnection, wire bonding and the via filling by ink-jet printing techniques are explored. Considering the need for identical sized die stacking, interconnections for the 3-layer stacked devices are demonstrated using ink-jet printing technique.

Chapter 6 summarizes the work done in this thesis. It also, discusses the future directions in a wide range of potential applications.

Chapter 2

Stacking the Ultra-Thin Silicon layers with functional MOS Devices and its reliability

This chapter presents a low temperature process to transfer devices on ultra-thin silicon layers from a parent substrate to a foreign substrate or stack. MOS devices were fabricated on SOI wafer. The device wafer was then temporarily bonded on a carrier wafer. The handle layer was etched and the remaining ultra-thin silicon device layer of ~1.4 µm was transferred to a foreign substrate using permanent bonding. Here, we explored two different bonding approaches, viz; (1) the gold-indium (Au-In) transient liquid phase (TLP) bonding and (2) the epoxy bonding. We demonstrate the advantages of epoxy bonding method over the TLP method. The unique characteristic of this epoxy bonding approach is its capability to vertically stack multiple thin silicon layers. Further, we demonstrate three-layer stacking of the ultra-thin silicon layers with functional MOSFET's in each layer. Electrical characterization results of NMOS/PMOS devices in each layer are presented before and after the transfer process and compared. The changes in measured device performance before and after stacking are studied using simulations. The maximum process temperature in this approach is 150 °C, which is considerably lower than those reported in the literature. This result demonstrates the feasibility of multi-layer low temperature silicon homogeneous stacking.

The work presented in this chapter is accepted recently with the following journal.
(1) NPV Krishna et. al, IEEE Transactions on Electron Devices, 2019.

Introduction

Three-dimensional (3D) stacking of the heterogenous components is a solution proposed to address design, manufacturability, power management, bandwidth and cooling challenges[55] associated with high performance computing and autonomous systems for Internet-of-Things (IOT)[5,54,56]. These requirements have driven the research towards development of various technologies like silicon interposer, embedded multi-die interconnect bridges[29], heterogeneous interconnect stitching technology [57] and multichip package[5,23,28,57,58]. Out of these aforementioned technologies, multichip packaging enables both homogenous and heterogenous integration[59] of the devices through stacking. Furthermore, this technology may enable the stacking and integration of MEMS[60], MOSFET[57,60,61], microfluidic[54,62,63], optical[64] and high power GaN based devices[65] with lowest form factor.

To address these requirements many researchers have reported ideas like 3D-stacking of ultrathin chip packages[66], 3D parallel layer processing[67] and multi-layer homogenous integration[21,67]. In all the above-mentioned cases, the silicon thickness was 10 μm and above[9,13,66–71]. However, there is always a need to reduce the silicon device thickness, which would significantly impact through-silicon-via (TSV) etch depths; interconnect lengths and metal/barrier fill requirements.

In this chapter we present a process flow, to transfer the ultra-thin silicon device layers, of thickness ~1.4 μm, to a foreign substrate, which enables vertical stacking of functional layers. Using the metal and the epoxy-based bonding methods, we demonstrate transfer of ultra-thin silicon layers having functional devices. TLP based metal bonding is a well-known technology in the MEMS packaging, where the alloy which is formed after bonding has higher melting point than one of the parent metals on both sides of the wafers[72-75]. Even though the epoxy-based bonding is a well-known idea for transfer of 50 μm thick wafer, in most of the cases the epoxy used to transfer was BCB[76] (Benzocyclebutene), wherein the thermal budgets were around 340 °C. On the other hand, in the present work we carry out the bonding processes using EPO-TEK 377 epoxy with a bonding temperature of maximum 150 °C.

2.1. Fabrication

2.1.1. Transistor fabrication NMOS and PMOS

Self-aligned n-channel and p-channel metal-oxide-semiconductor field effect transistors (MOSFET) were fabricated on p-type and n-type SOI wafers respectively with 2 µm thick device layer, 1 µm thick buried oxide (BOX) layer and handle layer of thickness 450 µm. The fabrication process flows utilized for both the NMOS and PMOS layers are briefly shown in

Figure 2. 1. To determine the diffusion schedules for the source / drain regions as well as to achieve the required sheet resistance for the contact regions, TCAD Athena simulations were performed. The device layer thickness post transistor fabrication was ~1.4 µm.

(a) NMOS process flow
- Field Oxide – Pyrogenic oxidation -1 µm
- Active area definition
- **Gate oxide** : dry oxidation 50 nm
- **LPCVD**: Poly-Si: 1000 nm
- Gate definition and Poly –Si etch
- **S/D Diffusion:** Phosphorus: 1000 °C 20min & 15 min and PSG removal
- Body contact definition and oxide etch
- Back Side Poly and oxide etch
- Al Metal contact definition and sputtering
- Forming Gas Anneal : 400 °C 20 min

(b) PMOS process flow
- Field Oxide – Pyrogenic oxidation -1 µm
- Body contact definition and oxide etch
- **Body Diffusion:** Phosphorus: 1000 °C 20min & 15 min and PSG glass removal
- Active area definition
- **Gate oxide** : dry oxidation 50 nm
- **LPCVD**: Poly-Si: 1000 nm
- Gate definition and Poly – Si etch
- **S/D Diffusion:** Boron : 900 °C 10 min & 1050 °C 30 min and BSG removal
- Body contact definition and Poly etch
- Back Side Poly and oxide etch
- Al Metal contact definition and sputtering
- Forming Gas Anneal : 400 °C 20 min

Figure 2. 1: Process flow for the NMOS and PMOS MOSFET fabrication

2.1.2. Ultra-thin silicon transfer using epoxy and Au-In TLP bonding

The above transistors were passivated with ~90 nm low temperature (140 °C) PECVD oxide, leaving only the contact pad areas open for interconnects, as shown in Figure 2. 2 (a) and Figure 2. 3 (a). The device side of the SOI substrate was temporarily bonded to a carrier glass substrate using temporary bonding material Brewer Science HT: 10:10 as shown in Figure 2. 2 (b) and Figure 2. 3 (b). During the temporary bonding process, the back side of the SOI substrate (i.e. handle layer) was protected using photoresist (PR). This was done to prevent contamination of the handle layer surface by HT: 10.10 residues, which was observed to interfere with the subsequent handle layer etch. The PR was removed and the handle layer of the SOI wafer was etched away using isotropic dry etch process in the deep reactive ion etching (DRIE) tool. The etch stopped at the BOX layer as seen in Figure 2. 2 (c) and Figure 2. 3 (c).

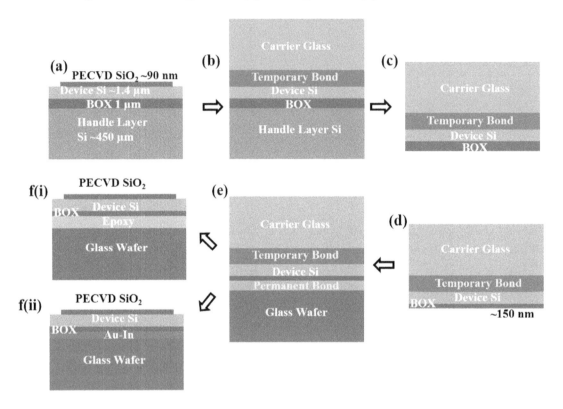

Figure 2. 2: Schematic of the process flow for the Ultra-Thin Silicon fabrication. (a) ~1.4 μm thick Si device layer after transistor fabrication. (b) Temporary bonding of glass carrier. (c) Removal of the SOI handle layer. (d) Controlled etch of BOX. (e) Permanent bonding of glass wafer. f(i) Final device with epoxy bonding and f(ii) Final device with Au-In bonding.

During further processing, cracks were observed on the thin device layer as shown in Figure 2. 3 (d). Further it was also noted that these cracks were present only for the transfer of 2 μm device layers. Similar transfer of 10 μm device layers did not result in such cracks. Therefore, these cracks were attributed to the stress arising from the deposited layers on the top and the BOX at the bottom of the device silicon layer. By sequentially removing the different layers from a dummy SOI wafer and measuring the curvature using the kSA MOS stress measurement tool, the stress of the buried oxide layer was found to be ~ 290.5 MPa compressive. After handle layer removal, the compressive

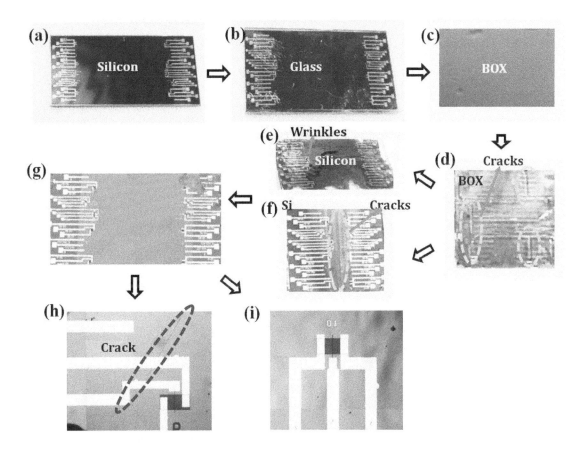

Figure 2. 3: (a) Top view of the as fabricated transistor SOI die. (b) Glass carrier wafer temporary bonding over the SOI die. (c) Image showing the BOX after handle layer removal using DRIE. (d) Stress cracks were observed after the transfer process upon further processing. (e) Wrinkles were observed on the transferred devices after complete removal of BOX. (f) Cracks were observed on the transferred devices with partial etch of the BOX. (g) Photograph of a decently bonded sample without any wrinkles or cracks. (h) Observed micro-cracks on a transistor after transferring to a foreign substrate using Au-In bonding. (i) Micrograph of the epoxy-bonded sample without any cracks.

BOX relaxes by stretching. This leads to cracking of the silicon device layer. Hence, it was concluded that the solution for reducing the residual stress effect is to reduce the thickness of the BOX layer. Removing the BOX layer completely, however, led to the wrinkles after transfer as shown in Figure 2. 3 (e). These wrinkles can be attributed to the remaining stress due to the bonding layer and the layers on the top side of the thin silicon device layer. Crack free and wrinkle free thin silicon, as shown in Figure 2. 2 (d) and Figure 2. 3 (g), could be achieved when a controlled wet etching of the buried oxide layer was carried out to reduce it to ~150 nm. Stress cracks are observed when more than 200 nm of BOX was left as shown in Figure 2. 3 (f). We observed that leaving between 100 - 200 nm BOX on the silicon device layer balances the stress induced by the bonding layer and the different layers present on the front side of the device layer.

The ultra-thin silicon device layer was transferred on to the permanent package substrate (e.g. glass), using two approaches as shown in Figure 2. 2 (e). For transfer, we explored: (I) room temperature or low temperature epoxy bonding (II) low temperature Au-In bonding and as described below:

2.1.2. 1. Method (I)

In this epoxy bonding approach, EPO-TEK UJ1190 was spin coated at 1600 rpm on to a cleaned glass substrate. The ultra-thin Si device layer on temporary substrate was brought in contact to the glass substrate and exposed to UV light using the MJB4 lithography tool for 60 seconds. The exposure was completed in 15 cycles at intervals of 30 seconds at room temperature. This resulted in a permanent bond between the ultra-thin silicon and the glass substrate as shown in Figure 2. 2 (f(i)). The carrier substrate was then separated using thermo-mechanical debonding on a hot plate at 150 °C as recommended by the manufacturer. The remaining temporary bonding material was thoroughly cleaned using the wafer bond remover from Brewer Science.

2.1.2. 2. Method (II)

For Au-In bond, Cr/Au (20/100 nm thickness) was sputtered on the back of ultra-thin silicon and 1 μm indium was evaporated on the foreign substrate (in this case glass substrate). These two substrates were brought together and subjected to 600 N force at a temperature of 150 °C for 30 min in a bonding tool under vacuum. Prior to the bonding, oxide was removed from the indium surface using a dilute HCl dip just. After the bonding process, the carrier wafer was separated from the stack by the thermo-mechanical debonding method. This was followed by cleaning of the remaining bonding material. The final device is shown in Figure 2. 2 (f(ii)).

From the SEM/FIB pictures in Figure 2. 4 (a), the post bonding thickness of Au-In alloy was 1.3 μm whereas, the epoxy thickness was ~25 μm as shown in Figure 2. 4 (b). For Au-In bond the SEM/FIB image clearly shows voids. The incomplete bonding is attributed to the low bonding temperature. Upon careful observation under the microscope, micro cracks on the ultra-thin silicon layer were observed at random location for the case of indium bonded sample as shown in Figure 2. 3 (h). These cracks could result from large bonding forces used in the bonding process. On the other hand, in the case of epoxy bonding cracks were absent (see Figure 2. 3 (i)) and the yield was almost 100 %. Absence of cracks for epoxy bonding can be expected due to lower stiffness of epoxy and lower bonding force. Further, the epoxy bonding process is relatively cheap compared to the TLP bonding which involves two metal deposition steps and specialized bond tool requirements.

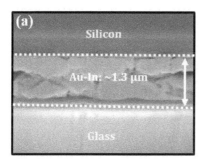 

Figure 2. 4: FIB images of the bond interface using (a) Gold Indium transient liquid phase bond with Au-In alloy thickness of ~1.3 µm (b) Epoxy bonded sample with epoxy thickness of ~25 µm.

2.2 Vertical Stacking

2.2.1. Process flow

Based on the better results obtained with the epoxy bonding, we adopted this approach for vertical stacking of multiple layers of ultra-thin silicon substrates. The stacking scheme remains the same as discussed in the previous section and is shown in Figure 2. 5. It involves temporary bonding of a die to the carrier wafer; backside silicon removal and controlled oxide etching followed by permanent bonding as shown in Figure 2. 5 (b) and (c). For further stacking, a SU-8 2002 layer was used to planarize the stack and provide dielectric isolation. Using a photolithography step, the contact pad areas were opened, all the other areas were left with ~1-2 µm thick SU-8. Before bonding the 2nd layer, SU-8 surface was treated with low power oxygen plasma. This step is required for SU-8 surface activation which allows proper spreading of epoxy and better adhesion.

The 2nd layer of ultra-thin Si device layer was prepared using the same technique as described above. Silicon being not transparent to the UV light, UJ1190 epoxy cannot be used for further stacking (i.e 2nd and 3rd layer). Hence, a thermally curable epoxy (EPO-TEK 377) was explored and used for further bonding/stacking. The selected epoxy has very low viscosity and low outgassing, both properties being beneficial for the stacking process. The epoxy was spin coated on to the backside (BOX) of the ultra-thin Si device layer at 1800 rpm for a minute. This was followed by bonding on the hot plate at 135 °C for 30 minutes. After bonding, the carrier wafer was removed using thermo-mechanical debonding method.

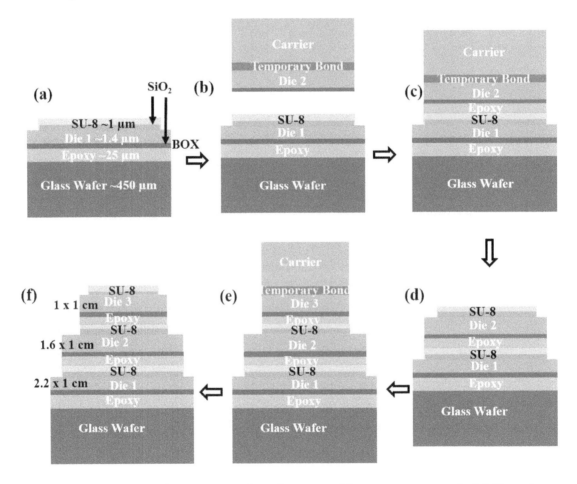

Figure 2. 5: Schematic of the process flow for vertical homogenous stack. (a) Transistor fabrication and transfer to the glass wafer using epoxy bonding methodology. (b) Preparation of second die for the transfer over the first one – Temporary bonding of carrier glass to SOI, etching of handle layer silicon and partial etch of oxide. (c) Permanent bonding of the device stack using epoxy. (d) Separation of the Glass carrier and thorough cleaning. (e) Another ultra-thin silicon permanent bonding to the existing stack. (f) Final stack.

In preparation for the transfer of 3rd layer, 1 μm thick SU-8 was spin coated on the stack and patterned over the device region as shown in Figure 2. 5 (d)). Finally, the third die was also bonded to the existing stack with devices facing to the top as shown in Figure 2. 5 (e). Thermo-mechanical debonding of the carrier wafer, lead to the completion of the stacking process as seen in Figure 2. 5 (f). Successful stacking was also achieved by bonding the first layer using thermally cured epoxy (EPO-TEK 377) instead of UV cured epoxy (UJ1190).

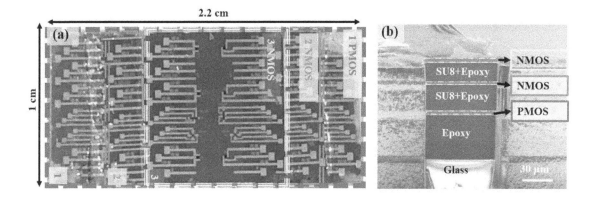

Figure 2. 6: Image of the 3-layer stack (a) Top view photograph showing all the three bonded layers. (b) Cross-sectional FIB image showing all the three functional NMOS and PMOS layers.

As seen in Figure 2. 5 (f) the three layers in the stack used dies of different dimensions as follows: Die 1 was 2.2 cm x 1 cm, die 2 was 1.6 cm x 1 cm, and Die 3 was 1 cm x 1 cm. The lengths of the dies were designed to be different to ensure that both sides of the die can be accessed from above to probe the device contact pads as seen in Figure 2. 6 (a). In the fabricated stack the bottom most layer consisted of PMOS devices, whereas the middle and the top layer consists of NMOS devices. Figure 2. 6 (a) shows the photograph of the stack, where the top third layer with devices and the interconnect pads on both sides of all the three layers can be seen. Figure 2. 6 (b) demonstrates the cross-sectional FIB image of the 3-layer stack with epoxy and silicon layers.

2.2.2. Keep out zone

FIB cross-sections at various positions on the stack can be seen in Figure 2. 7. Image in the left shows the edge of the two-layer stack and the image on the right shows after stacking the three layers. From the images it can be observed the effect of the epoxy on the edges of the bottom and the top die due to spin coating and bonding is ~ 40 μm. Hence, the keep out zone for the devices on the edges of the both the stacked silicon layers would be ~40 μm from the respective edge.

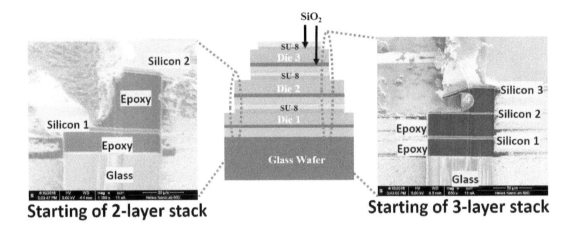

Figure 2. 7: Epoxy spread at the bond interfaces to the subsequent layers. Keep out zones in both the layers was ~ 40 μm.

2.3 Characterization of the transferred devices

2.3.1. Diffusion measurements

The PMOS and NMOS transistor (gate width = 200 μm, length = 20 μm) structures were simulated using TCAD Athena tool to identify the right diffusion parameters for fabricating the experimental devices. Based on the simulation study, phosphorus pre-deposition and drive-in both were performed at 1000 °C for 20 min and 15 min respectively, to realize the S/D regions in the NMOS transistor. For the PMOS, the boron pre-deposition was performed at 900 °C for 10 min followed by drive-in at 1050 °C for 30 min. The sheet resistance (SR) of fabricated NMOS and PMOS were ~4.6 Ω/□ and ~30 Ω/□ respectively.

2.3.2. DC electrical measurements

Electrical DC measurements of the NMOS transistors were performed before and after the transfer to foreign substrate using both the "Au-In" and "epoxy" bonding approaches. I_d-V_d (output) measurements were performed at V_{gs} = 0 to 7 V in steps of 1 V and V_d = 0 to 5 V. I_d-V_g (transfer) measurements were performed at V_{ds} = 7 V and V_{gs} = -2 to 5 V. Figure 2. 8 (a) and Figure 2. 8 (c) summarize the output characteristics of both the bonding approaches. It can be observed that at lower V_g before and after transfer, the Id values were identical in both the bonding cases. However, in both the cases, reduction in current can observed for V_{gs} >3V. This could be attributed to the poor thermal conductivity of the glass substrate. Also, it can be observed that reduction in current is lower in the case of Au-In bond as compared to epoxy bond at higher gate voltages; This can be attributed to better lateral dissipation of heat through the Au-In alloy. Whereas the epoxy being a poor conductor of heat this will lead to higher device temperatures and hence the observed current reduction.

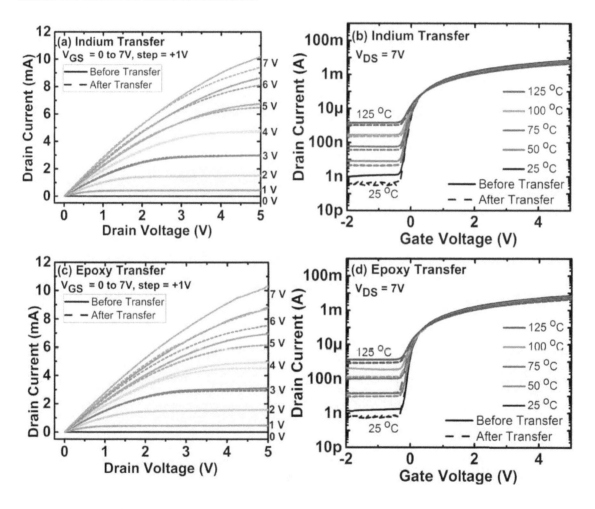

Figure 2. 8: Electrical characteristics of the 20 μm channel length NMOS device before and after transfer to foreign substrate. Indium bond (a) Output characteristics at room temperature and (b) Transfer Characteristics at various temperatures. Epoxy bond (c) Output characteristics at room temperature and (d) Transfer Characteristics at various temperatures

Figure 2. 8 (b) and (d) summarize the transfer characteristics (I_d versus V_g) at various temperatures (25 °C to 125 °C, in steps of 25 °C) respectively for the indium bonding and epoxy bonding cases. The off-current after the transfer are compared with those obtained before transfer for both cases. As expected, the off-currents are observed to increase with temperature. For both the bonding approaches, the transferred devices showed lower off-currents than the device on the SOI substrate. In the case of indium bonding, the extracted threshold voltage (V_t) before stacking was -86.01 mV and after stacking it was -67.4 mV. In the case of the epoxy bonding, the extracted V_t before stacking was -53.9 mV and after stacking it is equal to -43.3 mV. These results indicate about 20% shift in the threshold voltage for both the approaches.

2.3.3. Negative differential resistance

To confirm that the reduction in on-current is due to the temperature rise in both the bonding techniques, we carried out the I_d-V_d measurements at elevated temperature and up to higher drain voltages (V_d = 15 V) for V_g = 10 V on both Au-In and epoxy bonded NMOS device samples. These measurements were carried out between room temperature (RT) (25 °C) and 125 °C at an interval of 25 °C. Similar measurements were also carried out between 25 °C and 200 °C for NMOS devices fabricated on the SOI wafer and retained without transferring them on to foreign substrate (we designate them as the "Original NMOS SOI device"). Figure 2. 9 (a) and (b) show the I_d-V_d measurements of the NMOS transistor transferred to glass using Au-In and epoxy bonding approaches respectively. Figure 2. 9 (c) shows the results obtained on the original NMOS SOI device. As expected, in these devices, due to rise in the applied chuck temperature the drain current (saturation current) reduced at higher temperatures. The decrease in the saturation current occurs because of decrease in mobility of charge carriers in channel caused by increased electron-phonon scattering at higher temperatures. For the transferred substrates we also observe a region of negative differential resistance (NDR) region[77,78]. Very minimal effect of the same can be observed in the original NMOS SOI device as shown in Figure 2. 9 (c).

Once the MOSFETs are transferred to a glass substrate through Au-In or epoxy bonding, the effect of poor thermal dissipation on output characteristics becomes prominent, even at room temperatures. In both the cases, with increase of drain voltage, the drain current increases to a maximum and then starts to decrease leading to an NDR region as can be seen in Figure 2. 9 (a) and Figure 2. 9 (b). This happens as the power through the device increases the device channel heats up due to the poor thermal dissipation through the bonding material and the glass substrate. On comparing the MOSFET characteristics in case of Au-In bonding with those of epoxy bonding, the epoxy being an even poorer thermal conductor, the reduction in device current is further enhanced.

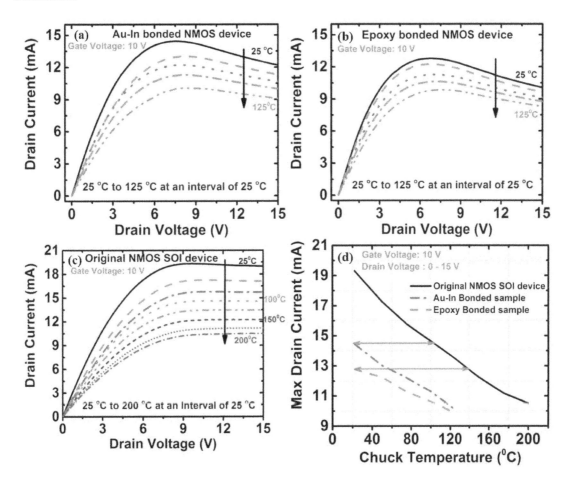

Figure 2. 9: Device behaviour at elevated temperatures (a) Output characteristics of the Au-In bonded sample (b) Output characteristics of the epoxy bonded sample (c) Output characteristics of the original NMOS SOI sample (d) Mapping the temperature of the Au-In bond and the Epoxy with respect to the as fabricated sample using the maximum current obtained.

To measure the rise in temperature, the maximum current obtained in all the three cases was recorded and plotted in Figure 2. 9 (d). It may be noted from the graph, that the maximum current for room temperature (RT) measurements in the Au-In bonded devices approximately corresponds to those obtained at 100 °C in the original NMOS SOI device. Whereas, for the epoxy bonding approach, the maximum current value at RT corresponds to ~ 150 °C in the original NMOS SOI device, as shown in Figure 2. 9 (d). Hence, the Au-In bonded sample would be at 100 °C and the epoxy sample would be 150 °C at the maximum current carrying conditions. As V_d increases further the power dissipation in the channel increases leading to further rise in temperature and hence, reduction in current with increase in V_d.

To understand the NDR effect on the post transfer devices, TCAD Atlas simulations were conducted on the NMOS SOI transistor. Steady state I_d-V_g simulations results at higher drain voltages on SOI wafer (with no vertical stacking) can be seen in Figure 2. 10 (a). Simulated measurements were identified to be in close match with the original fabricated NMOS SOI device measurements as shown in Figure 2. 10 (a). The Au-In and epoxy bonded devices results (after fabrication) were also plotted and at higher drain voltages where a significant negative differential resistance can be observed. The same has been verified through thermodynamic MOSFET simulations using TCAD Atlas as shown in the same figure. We suspect the significant NDR effect in the post transferred devices is due to the increase in the channel temperature during operation. This could be due to the lack of proper thermal dissipation through the substrate. To verify the same, we have simulated the increase in the channel temperature with increase of the drain voltages for all the devices at room temperature as shown in Figure 2. 10 (b).

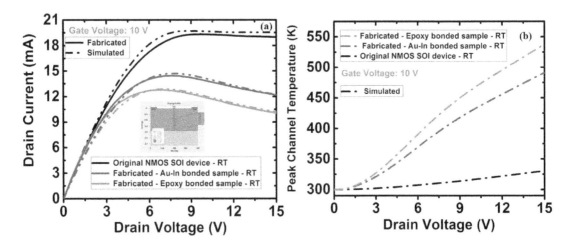

Figure 2. 10: (a) Comparison of the simulated and fabricated output characteristics of pre and post transfer of the NMOS SOI device to glass substrate using Au-In bond and the epoxy bond (b) Channel temperature simulation measurements with change in drain voltages, indicating significant channel temperature rise post transfer.

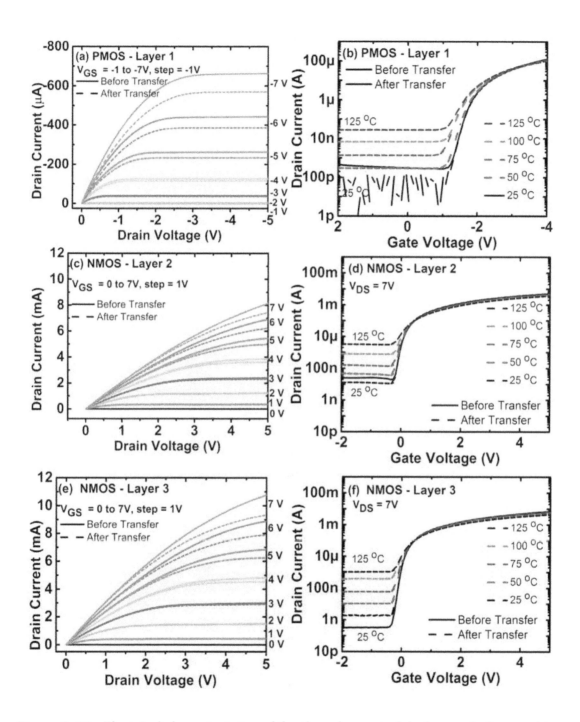

Figure 2. 11: Electrical characteristics of the three layer stack before and after stacking comparison (a) Output characteristics of the first/bottom (PMOS) layer in the stack (b) Transfer characteristics of the first/bottom (PMOS) layer in the stack at room temperature and elevated temperatures (c) Output characteristics of the second (NMOS) layer in the stack (d) Transfer characteristics of the second (NMOS) layer in the stack at room temperature and elevated temperatures (e) Output characteristics of the third/top (NMOS) layer in the stack (f) Transfer characteristics of the third/top (NMOS) layer in the stack at room temperature and elevated temperatures

2.3.4. 3-layer stack with functional devices and DC measurements

Having established the impact of both the bonding processes on device characteristics, due to the low temperature and cost-effectiveness of the epoxy bonding process it was further utilized for 3-layer stacking. Each of these pre-fabricated layers has NMOS or PMOS devices. The first layer with PMOS devices, output and the transfer characteristics before and after transfer can be seen in Figure 2. 11 (a) and (b). The second and the third NMOS layer results can be seen respectively in (i) Figure 2. 11 (c), (d) and (ii) Figure 2. 11 (e), (f). Also, the stack was subjected to elevated temperatures and recorded the transfer characteristics to understand the behaviour of the devices upon stacking. No significant changes have been observed in the device behaviour upon stacking the functional MOSFET layers.

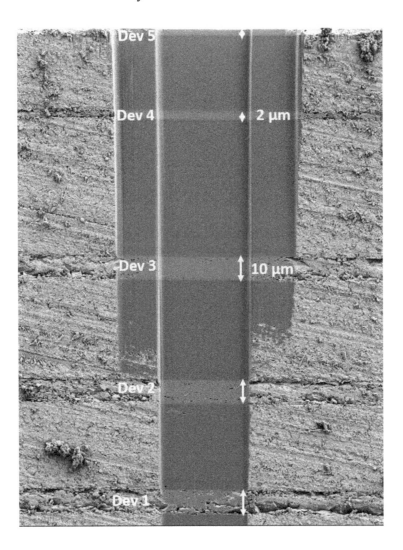

Figure 2. 12: Five-layer stack with functional ultra-thin si layers. First two layers of 2 μm thick and rest three layers of 10 μm thick.

2.3.5. 5-layer Ultra-thin silicon stack

In this section, we demonstrate the feasibility of extending this bonding process to a greater number of layers. Figure 2. 12 shows the FIB images of functional 5-layer stack with three 10 μm thick layers and two 2 μm thick layers. However, due to the step design, number of layers would be limited by the die size. To address this issue and to improve the silicon real estate, the die sizes can be replaced with equal sized dies incorporating metal filled TSV based interconnects. Silicon thickness being very low (~1.4 μm thick), the 'via' opening can be made smaller and lesser metal and barrier fills would be needed. However, the epoxy thickness is around 25 μm which is still to be reduced further to get the full benefit of this process.

2.4 Reliability measurements on the Ultra-Thin Silicon Stack

2.4.1. Fabrication of two-layer stacks with functional devices

Two-layer stacked devices were fabricated as shown in Figure 2. 13 using the process flow discussed in Section. 2.1.2.1. The first stack has PMOS devices in both the layers. Whereas, the second stack has NMOS in one and the PMOS in the other layer.

Figure 2. 13: Photograph of the 2-layer stacked die with layer 1 of 1.6 cm x 1 cm and layer 2 of 1 cm x 1 cm.

2.4.2. Reliability tests

The fabricated stacks were subjected to the industry standard Joint Electron Device Engineering Council (JEDEC) standard electrical reliability tests JESD22-A120A and JESD22-A104E-G. The first test JESD22-A120A for removal of moisture diffusivity involves exposure to 85 °C for 24 hours and for the water solubility and diffusivity the stacks were exposed to 85 °C with 60 % RH for 65 hours (However, the actual test is supposed to be done for 168 hrs). The second test (A104E-G) is a temperature cycling test to determine the device capability in withstanding extreme thermal shocks. This test involves sudden alternating low (-40 °C) and high (+125 °C) temperatures for 10 cycles each for a period of 30 minutes at each temperature. After subjecting to each of these tests the device behaviour has been tested by measuring the electrical DC characteristics of fabricated transistors.

2.4.3. DC electrical measurements

Fabricated NMOS and PMOS transistors before and after the transfer to glass substrate were electrically characterized using Agilent B1500 device analyser and compared as shown in Figure 2. 14 and Figure 2. 15. Identical devices were measured and compared after each stage of fabrication and the reliability tests. Figure 2. 14. shows a typical set of dc measurements of the two-layer stack with both the layers having PMOS devices.

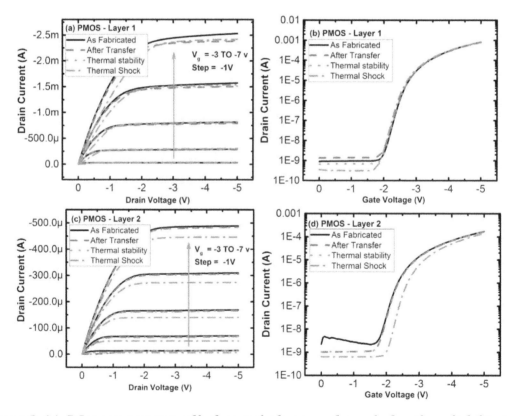

Figure 2. 14: DC measurements of before and after transfer and after the reliability tests (Thermal Stability - JESD22-A120A and Thermal shock - JESD22-A104E-G) of both PMOS devices in Layer 1 and 2. (a) Output characteristics with varying gate voltage for Layer 1 (b) Transfer characteristics for Layer 1 (c) Output characteristics with varying gate voltage for Layer 2 (d) Transfer characteristics for Layer 2.

The channel lengths of the fabricated PMOS Layer 1 and layer 2 were 6 μm and 20 μm respectively. The output characteristics with V_d = 0 to -5 V and V_g from -3 to -7 V in steps of -1 V and the transfer characteristics with V_d = 0 to -5 V were measured for both the layers. After the first part of the JESD22-A120A thermal stability (85 °C for 24 hours) test and the second JESD22-A104E-G test, the devices were characterized and compared as shown in Figure 2. 14. It may be noted that, after these intensive tests there is no significant change in the device behaviour. However, at higher gate voltages slight reduction in the drain current has been observed. This could be attributed to using glass as the permanent substrate, which is a poor thermal conductor.

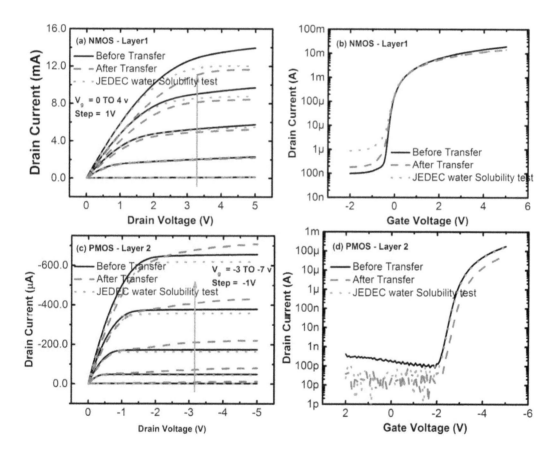

Figure 2. 15: DC measurements of before and after transfer and after the reliability tests (Solubility test - JESD22-A120A) of the NMOS device in Layer 1 and PMOS device in layer 2. (a) Output characteristics with varying Vg for Layer 1 (b) Transfer characteristics for Layer 1 (c) Output characteristics with varying Vg for Layer 2 (d) Transfer characteristics for Layer 2.

Images in the Figure 2. 15. shows another two-layer stack DC measurement with Layer 1 NMOS and the layer 2 PMOS devices. The NMOS and PMOS in Layer 1 and 2 are with channel lengths of 10 μm. The NMOS output characteristics with V_d = 0 to 5 V and V_g from 0 to 4 V in steps of 1 V and the transfer characteristics with V_d = -2 to 5 V were measured and presented in Figure 2. 15. The PMOS output characteristics with V_d = 0 to -5 V and V_g from -3 to -7 V in steps of -1 V and the transfer characteristics with V_d = 2 to -5 V were performed. After the second part of the JESD22-A120A (water solubility test - 85 °C with 60 % RH for 65 hours) test, the devices were measured and compared as shown in Figure 2. 15. After each of these intensive tests there it may be noted that there is no significant change in the device behaviour.

Conclusion

In summary, we have successfully demonstrated transfer of ultra-thin silicon layers to package substrate (glass) using epoxy and indium bonding-based transfer methodologies. Further, we have shown that even though the indium-based bonding approach displays better currents as compared to epoxy-based bonding approach, it suffers from difficulties associated with TLP bonding and reduced yield due to crack formation in device layers. On the other hand, epoxy bonding is easier and economical, with a higher yield after stacking, giving an advantage over TLP method. We have used epoxy bonding methodology to demonstrate a 3-layer ultra-thin silicon stacking process with functional transistors. The NMOS and PMOS devices on ultra-thin silicon layer have been characterized and compared at various stages of fabrication. Our experimental results exhibit minimal deviation in the post-stacking device behaviour. At higher drain voltages, significant reduction in current has been observed in both the epoxy and Au-In based bonding approaches. Based on our simulations and measurements, we conclude that the above reduction in current is due to the poor thermal-conduction properties of the foreign substrate. However, as this stacking process is a post device fabrication process; it can be performed in a backend foundry with no need of changing any front-end foundry processes. Also, the devices in this study have exhibited negligible change in performance after basic JEDEC reliability tests, indicating the robustness of the transfer process. Although glass is used as an example of a substrate here, the process outlined here can be replicated on other substrates to facilitate better heat extraction from the devices and/or enable conformal/flexible electronics. This novel simple process may pave way towards 3D stacked ultra-thin devices.

Chapter 3

Heterogenous Stacking of III-nitride-on-Si HEMTs and Si-MOSFETs to Copper

In this chapter, we demonstrate a new approach for stacking of AlGaN/GaN high electron mobility transistors (HEMTs) and Si MOSFETs onto a copper substrate. The main aim of this stacking approach is to develop a post fab vertical stacking methodology suitable for integration with the packaging industry. Here, both the GaN HEMT and MOSFET' devices were first fabricated on silicon substrate using separate process. Then GaN devices were epitaxially lifted off from the parent substrate and transferred onto a glass substrate using epoxy and Au-In bonding. However, due to poor thermally conducting property of the final substrate the device current was reduced. To address this issue, GaN devices were epitaxially lifted-off and transferred onto a copper substrate to improve thermal conductivity. However, large lattice and thermal mismatch makes transfer of GaN grown on silicon a non-trivial process. Thick electroplated copper was used to improve mechanical strength, allowing transfer of the GaN layer using Cu-In bonding. Next, an ultra-thin silicon layer (~1.5 μm) with functional NMOS transistors were separated from parent SOI wafer and then stacked above the GaN devices using cost-effective epoxy bonding approach. The Cu-In bonding not only improved thermal performance but also led to significantly better GaN device behaviour.

This GaN work is done in collaboration with Prof. Digbijoy Nath, Prof. Muralidharan and Prof. Vasu. The growth and the HEMT fabrication were performed by the collaborators.

> The work presented in this chapter is under review in the following journal.
> (1) PVK Nittala et. al, IEEE EDL, 2018.

Introduction

In the last decade, there have been significant improvements in processing technologies which have led to the heterogenous integration of GaN and CMOS devices[65,79-82]. These integration of devices from different fabrication platforms allow for further scaling and functional diversification, which are highly needed for the futuristic IOT based systems. One approach to achieve this heterogeneous integration is through pre-fabrication transfer of the III-V layer on the silicon substrate. The multilayer substrate thus obtained is then processed to fabricate devices on the different layers. Such an approach is however restrictive, due to the contamination issues arising from the III-V substrates. This limits processing of such multilayer substrates to special foundries only [83,84].

A less restrictive approach of post-fabrication transfer of GaN on sapphire has been reported earlier[85-93]. GaN on silicon is much more advantageous than GaN on other substrates due its lower substrate cost, scalability and compatibility with the existing processing technologies. However, GaN growth on silicon presents far more challenges in terms of stress management due to a large lattice and thermal mismatch between silicon and GaN. This implies GaN layer on silicon has much larger stresses and stress gradients when compared to GaN grown on sapphire. Hence, the transfer of GaN layer grown on silicon, after substrate removal is a huge challenge.

Before we developed the AlGaN/GaN layers transfer from parent silicon to copper substrate, we explored transferring to glass substrate using epoxy and transient liquid phase (TLP) based Au-In bonding[73] methods. However, due to the poor thermal conducting property of the epoxy and the glass substrate reduction in current has been observed after the transfer process. To resolve this issue and improve heat dissipation, we have developed a transfer process to copper.

Our transfer process involves a novel method of bonding, which uses a thick interfacial electroplated copper. The electroplated copper provides the mechanical strength required for the transfer of the highly stressed GaN layer. This copper is also found to improve the GaN HEMT characteristics including a higher on/off ratio after transfer, for instance. By this method we have successfully transferred GaN HEMT devices fabricated on silicon substrate on to a copper substrate. Further, MOS transistors fabricated on an SOI wafer were thinned, and then, stacked above the GaN devices using epoxy bonding process. This functional layer transfer process allows vertical stacking of heterogeneous devices. The proposed integration can be carried out in the backend foundry using pre-fabricated devices with very minimal modifications in the existing GaN or CMOS foundry processes.

3.1. Transfer to glass using epoxy bonding method

3.1.1. Growth and transistor fabrication AlGaN/GaN HEMT

AlGaN/GaN HEMT stacks were grown on <111> oriented 2-inch Si wafers of high resistivity (>10 K ohm-cm) in an Aixtron metal organic chemical vapor deposition (MOCVD) reactor. The precursor sources for Gallium, Nitrogen, and Aluminium were Trimethylgallium (TMGa), Ammonia (NH_3) and Trimethylaluminum (TMAl) respectively. Hydrogen was used as the carrier gas during the growth process. The stack consisted of 50 nm Aluminium Nitride (AlN) grown at 1050 °C, and 100 nm AlN grown at 1150 °C (40 mbar pressure) as buffer layers. This was followed by step graded AlGaN transition layers to reduce dislocation density, which includes 75 %, 50 % and 25 % NH_3 for 250 nm, 250 nm and 500 nm respectively at 40 mbar pressure. On top of the transition layer, GaN buffers were grown at following conditions: 500 nm of GaN at 1000 °C at 40 mbar pressure and 200 nm of GaN at 200 mbar pressure with 1 slm of NH_3 flow. Finally, the active HEMT stack comprising of 1 nm AlN interlayer, 22 nm of 25 % AlGaN and 3 nm of GaN cap layer were deposited at 1050 °C and 40 mbar pressure.

The transistor was fabricated with optical lithography followed by Ohmic contacts (Ti (20 nm)/Al (120 nm)/Ni (30 nm)/Au (50 nm)) using an e-beam evaporator, and lift-off process. Subsequently, devices were processed using rapid thermal annealing in N_2 ambient for 45 sec at 850 °C. Device isolation was performed using BCl_3 - Cl_2 chemistry in an inductively coupled plasma reactive ion etching tool (ICP-RIE). This was followed by gate lithography and Schottky gate contact metallization (Ni (20 nm)/Au (70 nm)) followed by annealing for 10 min in N_2 ambient at 400 °C. The fabricated transistors were of 7 µm gate length, 100 µm gate width and 17 µm source to drain distance. The gate to source spacing in these devices was 4 µm. The processed wafer was diced using automatic dicing tool to a die size of 1.6 cm x 1 cm. Further, these individual dies were used in the transfer process. The schematic of the die can be seen in Figure 3.1 (a).

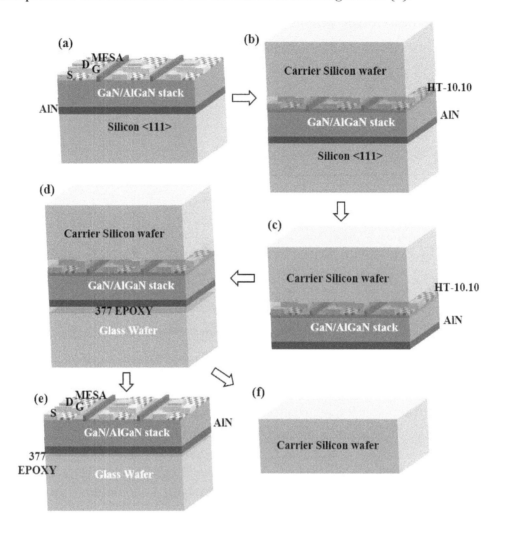

Figure 3. 1: Schematic of the process flow for the transfer of AlGaN/GaN HEMT from Si substrate to Glass. (a) HEMT's on silicon <111>. (b) Temporary bonding of the AlGaN/GaN HEMT to a carrier silicon wafer. (c) Isotropic etching of <111> silicon from the backside of GaN HEMT die. (d) Permanent bonding of the ultra-thin GaN/AlGaN layer to a glass substrate. (e) Separation of carrier wafer by mechanical debonding. (f) Carrier wafer available for the next wafer transfer process

3.1.2. Temporary bonding and silicon removal

Temporary bonding material HT:10.10 from Brewer science was spin coated on to the HEMT stack and bonded (Figure 3.1(b)) to a thoroughly cleaned silicon carrier wafer die, using the standard parameters provided by the supplier[94]. Back-side (<111>) silicon was next etched away (Figure 3.1(c)) using isotropic silicon etch recipe in the DRIE tool at an initial etch rate of ~15 μm/min for 15 min, followed by a slower etch rate of ~7 μm/min for 8 min. The reduction in the etch rate step ensures minimal stress on the device wafer die. The photograph after removing the silicon is shown in Figure 3.2 (a).

3.1.3. Permanent bonding using epoxy

EPO-TEK 377[95] was spin-coated at 1200 rpm on to a carrier glass wafer and permanently bonded to the thinned GaN die at 135 °C for 10 min on hotplate (Figure 3.1 (d)). The silicon carrier wafer was next de-bonded by the mechanical debonding method. Further, the AlGaN/GaN HEMT surface was thoroughly cleaned using the Brewer Science Wafer Bond remover solution to remove the excess HT:10.10 as shown in Figure 3.1 (e/f) and Figure 3.2 (b).

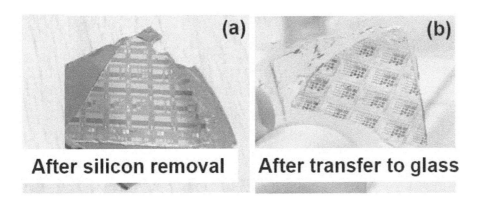

Figure 3. 2: (a) AlGaN/GaN die after silicon substrate removal. (b) After bonding to glass substrate.

3.1.4. DC electrical measurements

DC measurements were performed on the fabricated dies, to assess the electrical performance before and after transfer process. The results obtained by the electrical measurements were compared and shown in Figure 3.3 (a)and (b); No significant change in the transfer characteristics was observed in the after-transfer devices. After the transfer process the MESA leakage has been improved; Thich is attributed to the removal of parasitic conduction path through the Si[96,97]. However, the output characteristics was not satisfactory, a significant reduction in the current was observed after the transfer process as shown in Figure 3.4 (b). We believe, this is due to the poor thermally conducting property of the epoxy used for the bonding. Further, to understand the reliability of the transfer process, industry standard Joint Electron Device Engineering

Council (JEDEC) standard electrical reliability tests[98] were performed and the results are presented in the following section 3.1.5.

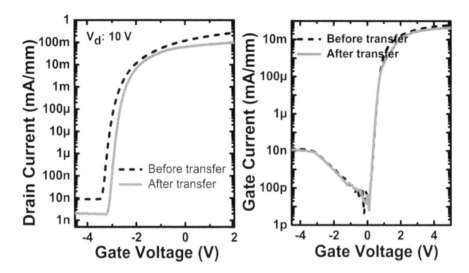

Figure 3. 3: DC electrical measurements: (a) Transfer characteristics before and after transfer on glass wafer. (b) Gate leakage characteristics before and after transfer on Glass wafer.

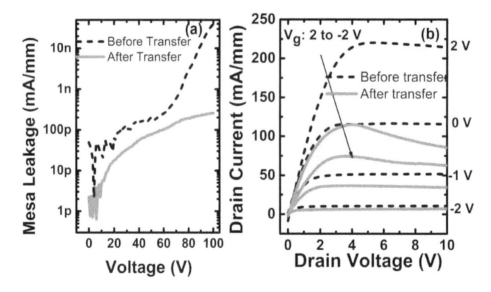

Figure 3. 4: (a) Mesa leakage (b) Output characteristics of the device

3.1.5. JEDEC reliability tests

The fabricated stacks were subjected to the industry standard Joint Electron Device Engineering Council (JEDEC) standard electrical reliability tests JESD22-A120A and JESD22-A104E-G. The first test (JESD22-A120A) is for the moisture diffusivity involving, exposure to 85 °C for 24 hours. The second test (A104E-G) is a temperature cycling test to determine the device capability in withstanding extreme thermal shocks. A104E-G involves sudden alternating low (-40 °C) and high (+125 °C) temperatures for 10 cycles each for a period of 30 minutes at each temperature. After subjecting to each of these tests, device behaviour has been tested by measuring the electrical DC characteristics of fabricated transistors. Figure 3.5 (a) shows the comparison of I_d-V_g characteristics of the AlGaN/GaN HEMT on silicon before and after the reliability tests. The plots indicate that there is no shift in the threshold voltage after the reliability tests. However, there is a slight increase in the off-current which can be attributed to the degradation of the buffer during the reliability tests due to the buffer traps[99]. Figure 3.5 (b) and Figure 3.5 (c) compare the transfer characteristics and gate leakage on the transferred devices. These results indicate that there are no significant changes in the device characteristics even after the JEDEC tests.

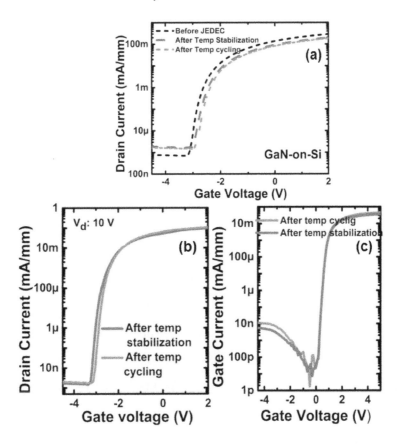

Figure 3. 5: JEDEC Measurements: (a) Transfer characteristics before transfer on Si wafer. (b) Transfer characteristics after transfer on Glass wafer. (c) Gate leakage characteristics after transfer on Glass wafer

3.2. Transfer to glass using TLP Au-In bonding method

3.2.1. Permanent bonding using Au-In bond

Here, for the permanent bonding, instead of the epoxy, TLP based Au-In bonding was employed. Using the same process discussed in the previous section 3.1.2, another set of devices were fabricated. In this process, the steps (a) and (b) shown in Figure 3.1 were first followed. Next, on the backside of the AlGaN die, Cr/Au of 20/100 nm thickness was sputtered and, on the glass die 1 µm thick Indium was evaporated. Further, these two substrates were brought together and subjected to 500N force at 160 °C for 30 min in a bonding tool under vacuum. After the bonding process, the temporary carrier wafer was separated using the mechanical debonding method.

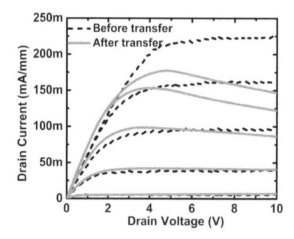

Figure 3. 6: Au-In TLP bond: GaN HEMT electrical measurements before and after transfer onto the glass substrate

3.2.2. DC electrical measurements

On the fabricated dies electrical measurements were performed to determine the device behaviour before and after the transfer process. Figure. 3.6 shows the output characteristics of the device. As compared to the epoxy transfer results shown in Figure. 3.4(b), the drain current improved from 100 to 170 mA/mm. However, it can be observed that there is some decrease in current after the transfer process. We believe this slight improvement in the current in the Au-In bonding approach is due to lateral dissipation of heat in the Au-In alloy. However, due to the poor thermal conducting property of the glass substrate and the rise in the channel temperature, current is reducing. To address this issue the AlGaN layers are to be transferred to the thermally conducting substrate as explained in subsequent sections.

3.3. Transfer to copper using TLP Cu-In bonding method

3.3.1. Growth and transistor fabrication HEMT and NMOS

3.3.1.1. AlGaN/GaN High Electron Mobility Transistor

AlGaN/GaN epilayer stacks were grown on <111> oriented 6-inch silicon wafer in an Aixtron close-coupled shower head system. A schematic of the GaN HEMT epi-stack is shown in Figure 3.7 (a). III-nitride devices were fabricated using conventional GaN HEMT processing steps. It involved forming Ohmic contacts using Ti (20 nm)/Al (120 nm)/Ni (30 nm)/Au (50 nm) deposition followed by rapid thermal annealing in Nitrogen (N_2) ambient at 850 °C for 45 s. Device isolation was performed using BCl_3 - Cl_2 based dry etching. This was followed by gate lithography and Schottky contact metallization Ni (20 nm)/Au (70 nm). Finally, the gate was annealed for 10 min in N_2 ambient at 400°C.

3.3.1.2. NMOS Transistor

Silicon MOSFETs were fabricated on p-type SOI wafers with 2 μm device layer, 1 μm thick buried oxide and 450 μm thick handle layer. Self-aligned n-channel metal-oxide-semiconductor field effect transistors (MOSFET) were fabricated using the process flow described in Sec. 2.1.1 (previous chapter). The process flow involves high temperature phosphorus diffusion, LPCVD deposition of 1 μm gate poly-si and dry thermal oxidation of the 50 nm gate oxide. The final device layer thickness after device processing was ~1.45 μm.

3.3.2. Temporary bonding and silicon removal

The device side of the GaN and MOS devices were separately bonded onto glass carrier wafers using temporary bonding material HT-10.10 from Brewer science. After the temporary bonding, back side silicon in both the cases was etched away in DRIE at an etch rate of ~14 μm/min. To ensure smooth landing on the AlN and SiO_2 for the GaN HEMT and Si MOS devices respectively, last few microns were etched at an etch rate of ~5 μm/min. On the AlN side (back side) of the GaN wafer, Cr/Au with 20/100 nm thickness was deposited using the sputter tool.

Further, copper was electroplated to a thickness of ~50 µm using the standard copper electro-plating processes. The thick electroplated copper provided mechanical strength to the GaN layer during the subsequent bonding and transfer. In absence of the electroplated copper the GaN layer was found to wrinkle and crack after the transfer process. This cracking was attributed to the large stress gradients existing in the GaN stack grown on silicon.

3.3.3. Permanent bonding using Cu-In bond to Copper

3.3.2.1. GaN layer transfer to copper plate

The copper plate used in this work was ~ 2 mm thick. It was electropolished [100] to reduce the surface roughness from ~20 µm to ~190 nm as revealed by AFM surface roughness measurements. 1 µm indium was evaporated on the polished surface using e-beam. The electroplated side of the GaN device was adhered to copper plate using molten indium at 200 °C for 10 min on a hot plate. After this, the temperature was reduced to 155 °C, to allow bonding for another 70 min. As explained in ref [101], at this temperature, Cu-In forms an alloy with a melting point of ~450 °C. It is important to note that during the initial phase of bonding both interfaces (temporary bonding material HT 10:10 and Cu-In) were in a liquified/soften state. This allows formation of a complete bond even in presence of surface bows. The transient liquid state also implies that in absence of sufficient mechanical strength the stressed GaN layer can crack. At the end of the bond time, carrier glass substrate was then separated using the mechanical de-bonding method and the remaining temporary bonding material was thoroughly cleaned using the wafer bond remover solution from the Brewer Science.

3.3.2.2. MOS stacking on to the GaN on copper

On the back side of the etched SOI wafer, SU-8 2002 at 6000 rpm was spin coated followed by the hard baking at 135 °C for 10 min. Further the SU-8 surface was treated with a low power oxygen plasma. This plasma treatment is required for surface activation, which allows for a uniform spreading of epoxy and improves adhesion. EPO-TEK 377 epoxy was spin coated on to the plasma treated surface at 4000 rpm, to provide a thickness of ~15 µm after bonding. This is followed by bonding the MOS layer onto the GaN for 45 min at 135 °C. Further the glass carrier was then separated using the mechanical de-bonding method. A schematic of final stack details can be seen in the Figure 3.7 (b) and the final device can be seen in Figure 3.7 (c).

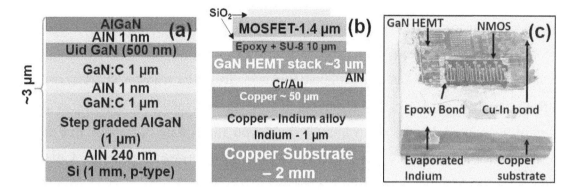

Figure 3. 7: (a) GaN HEMT stack used in this work (b) The conceptual image of the heterogeneously stacked GaN and MOS (c) Final device

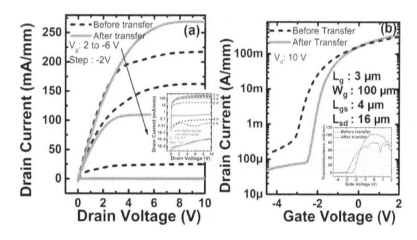

Figure 3. 8: GaN HEMT electrical measurements before and after transfer onto copper plate (a) Output characteristics, inset to the picture shows the log plot (b) Transfer characteristics

3.3.4. DC electrical measurements

The electrical DC measurements of the AlGaN/GaN HEMTs before the release from the parent substrate and after substrate transfer to copper plate are shown in Figure. 3.8. For the output characteristics, drain bias V_{ds} was swept from 0 to 10 V and gate bias V_{gs} was varied from +2 to -6 V in step of -2 V as shown in Figure. 3.8 (a). Before transfer, at 2V of V_{gs} the drain current density was ~220 mA/mm and after the transfer it has improved to ~270 mA/mm. At 0 V of V_{gs}, current density decreased from 160 to 110 mA/mm after transfer to copper plate. We also observe an improvement in the saturation after transfer to copper plate. For the transfer characteristics Vgs was swept from -4.5 V to 2 V at a constant V_{ds} of 10V as shown in Figure. 3.8 (b). Here, decent improvement in the on-off ratio can be observed after the transfer process. The inset to Figure. 3.8 (b) shows the transconductance plot with 80 mS/mm before and 103 mS/mm after the transfer to copper plate at V_{ds} of 10V. This indicates a significant improvement in the device behavior after transfer to copper. However, from the transfer characteristics a shift in threshold voltage can be observed, but it is not clear whether this shift is due to the strain induced by the electroplated copper on the AlGaN barrier or due to the exposure to the temporary bonding material during the transfer process. A detailed analysis is required to understand it better.

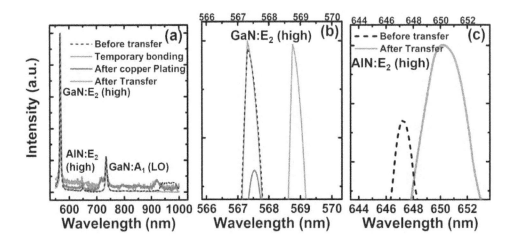

Figure 3. 9: Raman shift spectra for GaN device layers before and after transfer to copper plate (a) Full spectrum (b) GaN: E_2 peak shift (c) AlN: E_2 peak shift

To assess the strain in the bonded devices, Raman scattering measurements were carried out at various stages of the transfer process. Figure. 3.9 (a) shows the Raman data on the as fabricated HEMT device (before transfer), after removal of back side silicon, after copper electroplating and the final device bonded to copper plate. Figure. 3.9 (b) shows the GaN:E_2 (high) peaks with ~ +2 cm^{-1} shift indicating compressive stress after bonding to copper plate device. The AlN:E_2 (high) also has a shift of +3 cm^{-1} indicating compressive stress as shown in Figure. 3.9 (c). As explained in ref [102], shift in GaN:E_2 originates from GaN buffer and AlN:E_2 originates from AlN and Al$_x$Ga$_{1-x}$N intermediate layers. Hence, we believe that the enhancement in the device behaviour is due to the compressive stress induced by the plated copper and further Cu-In bond.

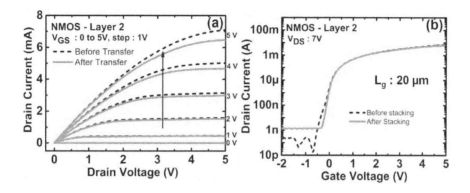

Figure 3. 10: NMOSFET electrical measurements of the before and after bonding to GaN device using epoxy (a) Output characteristics (b) Transfer characteristics

NMOS devices stacked above the GaN layer as shown in Figure 3.7 (b) and Figure 3. 11 were electrically tested and plotted in Figure 3.10 (a) and (b). MOS devices are low power devices and hence, were not affected by heat dissipation issues. So, for the transfer process, cost effective and simple epoxy bonding approach was used. From the results it is very clear that before and after bonding there is no significant change in the device behaviour. It can be observed that at higher gate voltages there is a slight reduction in the drain current, which could be attributed to the heat dissipation in the device or the transfer process itself. A minimal increase in the off-current has been observed for the "after-stacking" devices (Figure 3.10 (b)). Even though it's not very clear, it could be due to the extensive wet processing involved during the temporary bonding and transfer process.

3.3.5. Effect of die transfer to copper on the device temperature

Further, to understand the improvement in thermal performance due to the transfer, we measured the temperature rise as a function of power dissipation using an IR camera. Boron doped silicon ~1.8 kΩ resistors were fabricated on a 10 µm thick device layer SOI wafer. The device layer (10 µm thick) was separated and transferred to copper using Cu-In bonding method as shown in schematic Figure 3.11 (b). For comparison, the whole die (450 µm thick) was directly bonded on to a copper plate using silver epoxy as shown in schematic Figure 3.11 (a). Figure 3.12 shows the power dissipation results of a resistor on thick silicon (450 µm) and ultra-thin silicon bonded to copper plate. It is very clear from the graph that the transfer to copper plate has improved the heat dissipation. Inset to the Figure 3.12 shows the IR camera image.

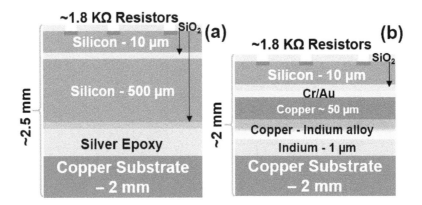

Figure 3. 11: Schematic of the stacks used for the thermal measurements (a) As fabricated SOI die (b) Transferred onto copper

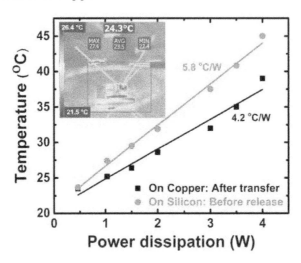

Figure 3. 12: Power dissipation of the resistors: before and after transfer onto copper.

Conclusion

GaN HEMT and MOSFET devices were fabricated and the active device layers were successfully transferred and stacked over a copper substrate. Using a thick electroplated Cu layer, we are able to mitigate the cracking issues arising from the large stress gradients existing in the GaN layer grown on Si substrate. The demonstrated Cu-In approach improved the GaN HEMT device behaviour due the strain induced by the copper electroplating. Very minimal reduction in current has been observed in the case of MOS devices after transfer. This easy and scalable heterogenous stacking of Si-CMOS and III-nitride devices enables 3D Integration of hybrid systems with more functionality, better power, smaller form factor and latency requirements.

Chapter 4
3D Integration of Heterogeneous Dies for Fluorescent Detection

Scaling by 3D integration of various heterogenous components enables miniaturized systems. However, heterogenous system integration is challenging due to the dissimilarities in materials and process used in fabrication of individual components. In this chapter we demonstrate a simple 3D integration method for miniaturisation of systems. Various components of the system were stacked using SU-8 based planarization and epoxy-based bonding. Spacer dielectric (SU-8) was patterned using photolithography for formation of interconnect vias. Electrical interconnects over the large topography between the layers was formed by screen-printing of silver nanoparticle epoxy. Using this integration technique, we demonstrate a fluorescence sensing platform consisting of silicon photodetector, plastic optical filters, commercial LED and a glass microheater chip. This chapter resolves several fabrication challenges of planarization, stacking and interconnection of these divergent chips. For example, process incompatibility of the plastic optical filters was resolved by additional passivation using parylene-C. The functionality of the demonstrated system is verified by detecting the fluorescence property of Rhodamine B and Rhodamine 6G dyes. Rhodamine B's sensitivity to temperature was also demonstrated using the on-chip microheater. This process flow can be scaled to stack a larger number of layers for demonstrations of more complicated systems with enhanced functionality and applications.

The work presented in this chapter has been published in the following journal.
(1) PVK Nittala et. al, ***IEEE Journal of Micro Electro Mechanical Systems (JMEMS)***, 2018.

Introduction

There have been reports[34,42,103] and products on the silicon based 3D integration of homogeneous memory and logic devices[29]. In contrast to homogeneous integration[67], building 3D systems by stacking the heterogeneous dies of MOS[57,64], MEMS[60], microfluidics[54], optical devices[64], etc. is significantly more difficult. Heterogeneous devices are fabricated on different substrates using different processing technologies. Hence, their integration into a single 3D stack requires addressing fundamental issues related to processing compatibility[104], different thermal budgets[105], different coefficients of thermal expansion and other differences in their mechanical properties. Also, technologies developed for 3D integration of CMOS devices are not easily translatable for 3D integration of Heterogeneous devices. For example, Through Substrate Via (TSV) technology is still not mature for non-silicon substrates and hence using TSV for 3D integration of heterogeneous dies is currently not feasible. Further, high temperature metal-based bonding techniques[72] cannot be used for integration of temperature sensitive components such as plastics.

We present an experimental prototype of a heterogeneously integrated system, which integrates multiple technologies to realize a 3D system for fluorescence detection application. From laboratory based research systems to the personal point-of-care diagnostic systems, microfluidics has been the enabling technology[106-111]. These lab-on-a-chip (LOC) systems have potential to replace large laboratory instruments and hence would make healthcare more affordable and accessible. Researchers have reported methods for packaging pre-fabricated CMOS dies which allows further processing and integration of microfluidic LOC platforms[21,54,62,104,112,113]. This has enabled integration of CMOS chips with sensing platforms on PCB or a chip carrier. Recently there has been a lot of research activity on integration of CMOS processed devices in flexible platforms[114], which will enable portable and wearable biosensing LOC diagnostics systems[54,69,105]. In these demonstrations interconnections were, however, made by either wire bonding or the flip chip bonding mechanisms which limits their capability to further stack other devices. On the contrary, our approach discussed in previous chapters allow us to integrate and interconnect a larger number of layers.

In this work we demonstrate a 3D stack consisting of a silicon device, two plastic filters, one LED and a glass microheater device. Various complementary metal oxide semiconductor (CMOS) based systems were reviewed in[54,110-113,115], with electrical, electrochemical, optical, thermal and magnetic sensing capabilities. In addition, optical filtering mechanisms have been discussed in detail in[116]. Polymer based filters were also explored[117]. Out of the optical, electrochemical and micromechanical detection mechanisms in microfluidic systems, fluorescence based bio sensing[107,118-124] is more prevalent due to its ease of implementation. The main benefits of this would be high selectivity, very low detection boundaries and availability of markers for tagging the bio markers in the bio-sensing platforms.

In this study, we present a cost-effective method of integrating heterogenous components in 3D stack using SU-8[125], by integrating diverse components with the aim of extreme miniaturization of a Spectro-fluorometer. A Spectro-fluorometer consists of (i) an excitation source, (ii) a detector and (iii) filters. Filters block wavelengths associated with excitation spectrum while allowing only the fluoresced light from the test sample to reach the detector. Further, to be able to perform chemical reactions like Polymerase Chain Reaction[126,127] (PCR) a microheater is required. In this chapter we report a detailed fabrication flow for heterogeneous stacking of a silicon-based photodetector, two optical plastic filters, a tiny commercial blue LED and a glass-based chip with microheater and interconnect pads. Using the developed heterogeneously integrated system in this work, we demonstrate temperature dependent fluorescence detection from Rhodamine B and Rhodamine 6G.

4.1. Individual components fabrication

The process flow developed for 3D integration of heterogeneous dies is shown in Figure. 4.1. In summary, the photodetector was fabricated on a p-type silicon substrate as shown in Figure. 4.1 (a). The details of design and fabrication of the photodetector are presented in the following section 4.1.1. Connections to the external world were planned on one side for all the dies (see Figure. 4.1 (b)). Commercially available plastic optical absorption filters were used in this fluorescence detection system. The filters were aligned and bonded manually over the photodetector (see Figure. 4.1 (b)). After bonding the filters, the stack was planarized using spin coated polymer SU-8 to enable further stacking as seen in Figure. 4.1 (c). Platinum resistive heater and interconnections for the LED were designed and fabricated on a thin glass substrate (150 µm thick). This glass substrate was aligned and bonded using epoxy over the stack (Figure. 4.1 (d)). After bonding, the stack was again planarized using spin coated polymer. The planarizing polymers were patterned using a lithography step to open vias for interconnections as shown in Figure. 4.1 (e). These vias were subsequently filled with silver epoxy using the screen-printing technique which enabled the electrical interconnection between the top glass wafer to the bottom silicon. Finally, a commercial blue LED was bonded over the glass wafer using silver epoxy (as shown in Figure. 4.1 (f)). Manually coloured SEM image of the device with all the components can be seen in Figure. 4.2. The demonstrated system has been scaled to the size of a USB stick as shown in Figure. 4.3. The process flow summarised above involves three main challenges: stacking, planarization and formation of interconnects. The fabrication details of the photo detector and glass fluidic chip with micro heater are presented in the following sections 4.1.1, 4.1.2 and 4.1.3.

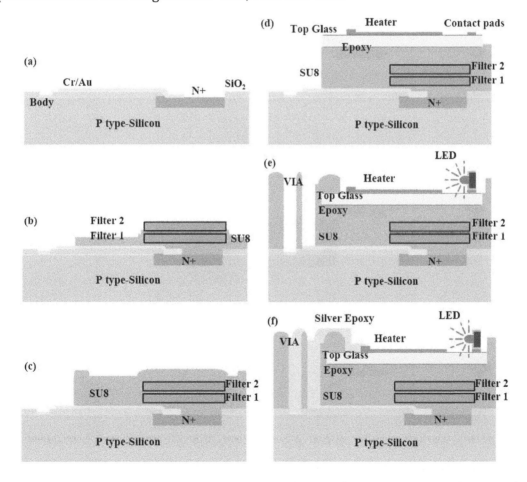

Figure 4. 1: Schematic of the process flow for 3D heterogenous stack. (a) Silicon photodetector fabrication. (b) Filter bonding using SU-8. (c) Planarization of the filter stack. (d) Bonding of the microfluidic chip (microheater on glass) using epoxy. (e) Planarization and via opening for interconnects. (f) Screen printed interconnects.

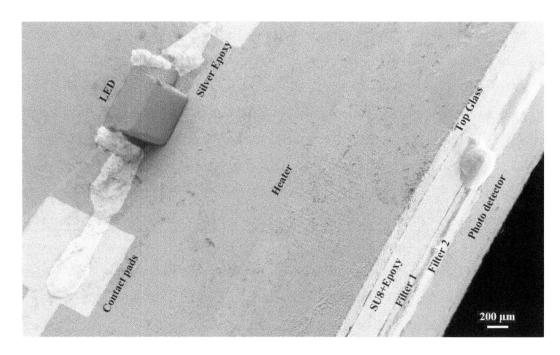

Figure 4. 2: False coloured SEM image of the device showing all the components in the system.

Figure 4. 3: Size of the proposed heterogenous system in comparison with a commercial USB stick

4.1.1. Photodetector Fabrication

In this work a simple silicon p-n junction photodetector was simulated and fabricated for fluorescence detection. The choice of the junction depth for the fabricated detector was based on the light penetration depth calculated from the optical absorption coefficient of the silicon as a function of wavelength[128]. From the intrinsic absorption coefficient for silicon as a function of wavelength[128], we estimate that the photons of

~500 nm wavelength will be absorbed beyond a depth of ~200 nm in the silicon. Hence, a shallow junction photodetector with a junction depth of 200 nm was designed using process simulator TCAD Athena. Even though this work is based on a simple photodetector, for better sensitivity this photodetector can be replaced with an avalanche photodetector without any significant modifications in the integration process.

Silicon photodetector die fabrication involves three lithography steps. In this fabrication scheme, 1 µm thick pyrogenic silicon dioxide was grown on a thoroughly cleaned 1-10 Ω-cm P-type <100> silicon wafer. The oxide diffusion mask was defined using lithography and oxide wet etch. During wet etching of oxide, backside of the wafer was protected using photoresist. Phosphorus (n+) diffusion was performed in a $POCl_3$ diffusion furnace at 900° C for 15 min (pre-deposition), followed by 10 min of annealing in the nitrogen ambient (drive-in) without removing the Phospho-Silicate Glass (PSG), which was formed during the pre-deposition step. After the diffusion process, PSG was removed in a dilute HF bath. In comparison with the simulated photodetector sheet resistance 16.14 Ω/□, the measured sheet resistance after fabrication was 17.33 Ω/□. The sheet resistance measurements were performed using the 4-probe measurement technique. This was followed by contact metallization using 20 nm chrome and 100 nm gold. Finally, after the metal patterning, the wafers were annealed in the forming gas ambient for 15 min at 400 °C to passivate the dangling bonds and interface states. The Fabricated photodetector die can be seen in Figure. 4.4 (a).

4.1.2. Selection of optical filter

As explained in previous report[116], an ideal filter for fluorescence detection application blocks (i.e. 0% transmission) the excitation wavelengths but transmits 100% of the emitted wavelengths allowing only the fluorescent light to reach the detector. The selection of the filter was based on the excitation and emission wavelength of the fluorescent dye and the radiant intensity of the LED. In this work blue LED with a peak wavelength of 430 nm was used as an excitation source. The excitation peak for the Rhodamine B (for fluorescence from Sigma Aldrich) used was 553 nm[129]. The filter was chosen to block the light having wavelength below ~ 550 nm as the emission peak for the dye used was 627 nm. Hence, this fluoresced light (627 nm) will be *detected* by the photodetector, and the filters will *block* anything below 550 nm. For this application commercially available 105 Orange filter from LEE Filters was chosen. Attaining lossless transmission and perfect blocking is however very difficult. Based on the attenuation of the blocked wavelengths, we decided to stack two filters to achieve sufficient rejection of the excitation LED.

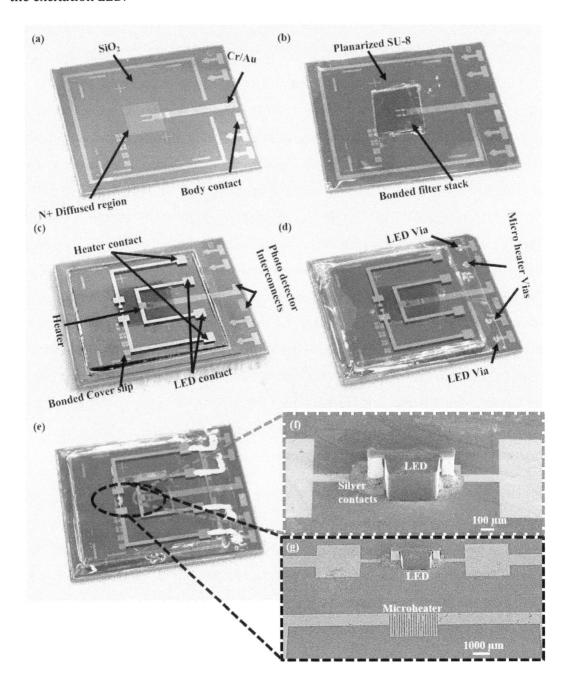

Figure 4. 4: Micrographs at various stages of the fabrication process. (a) Silicon photo detector. (b) Filters bonding and SU-8 planarization. (c) Glass microfluidic chip (with micro heater) bonding using epoxy. (d) Planarized stack and opened vias for metallization. (e) Final device with bonded LED and interconnections. (f) SEM image of LED and the Silver epoxy used for bonding LED. (g) SEM image of microheater and LED on the microfluidic chip.

4.1.3. Fabrication of the glass fluidic chip with microheater

A 2 kΩ platinum resistive heater was designed and fabricated on the ~150 μm thick glass cover slip. Piranha cleaned cover slips were patterned lithographically followed by the sputtering of Ti/Pt and photoresist lift-off in acetone to form the heater. On the coverslip along with resistive heater, bond pads for the excitation LED were designed as shown in Figure. 4.4 (c). The dimensions of the LED used in this work (from Kingbright) were 1.6 mm x 0.8 mm x 1.1 mm (thick). It is a blue colour GaN LED with a peak wavelength of 430 nm. Rather than mounting the LED horizontally with the LED facing upwards, it was mounted facing sideways as shown in SEM image Figure. 4.4 (f). This was done to illuminate the biological sample droplet which will reside in the same horizontal level. The LED was bonded manually, viewing under the microscope using silver epoxy as shown in Figure. 4.4 (f). The SEM mage of the microheater can be seen in Figure. 4.4 (g)

4.2. Hybrid Integration

4.2.1. Device Stacking by epoxy bonding

Stacking these heterogenous components in 3D and interconnecting them electrically was the main objective of this work. As the individual dies of the 3D stack were on heterogeneous substrates using through substrate via (TSV) was not feasible. Instead our approach uses interconnects at the edge of the dies as presented in previous chapters. To enable formation of electrical routing and interconnects at the edges, the stack needs to be planarized after each bonding step. For planarization spin coated SU-8 was used in this work. SU-8 is available in various ranges of viscosities allowing formation of planarizing layers with varying thickness. This allows stacking and planarization of devices having different thickness without significant changes in the process. An additional benefit of choosing SU-8 was its ability to be lithographically patterned even for very thick layers which makes formation of high aspect ratio vias for interconnects very simple.

Various epoxies including SU-8, EPO-TEK UJ1190, 377, 353-ND, Fevicol Fevikwik and M-Bond were investigated for bonding the components to the stack. The key properties of these bonding materials which effect the stacking are coefficient of thermal expansion (CTE), viscosity, glass transition temperature, out gassing, adhesion, curing methodology, colour (transparent or opaque) and optical properties. Properties of the bonding epoxies are compared in Table 4.1. Considering the low viscosity, very low out gassing and better optical transmission properties, EPO-TEK 377 and UJ1190 (shown inside the dotted line box in the table) inside the were selected for bonding of the microfluidic device. To keep the process simple SU-8 was used to bond the filters on the photodetector. The filters were aligned manually using the alignment marks structured on the photodiode chip. The alignment marks (+) can be seen in Figure. 4.4 (a) and Figure. 4.4 (b).

#	CTE 10^{-6} in/in°C Below Tg	Viscosity CentiPoise	Tg Glass transition temp °C	Outgassing @ 200°C	Adhesion kg/cm^2	Curable	Appearance	Comments	Optical properties @ 23°C
Fevicol-Fevikwik	-	40-50	-	High	~70	RT	Clear	Non-patternable	-
M-Bond	-	2.038	-	High	~2-2.8	RT	Amber	Non-patternable	-
EPO-TEK 353 ND	54	3000-5000 @ 50 rpm	≥ 90	0.22 %	> 140.6	Thermal	Amber/Dark Red	Non-patternable	≥ 50% @ 550nm and ≥ 98% 800-1000nm
EPO-TEK UJ1190	68	501 @ 100 rpm	100	0.04 %	-	UV	Pale Yellow	Non-patternable	≥ 94% @ 520nm -1000 nm and ≥ 80 % 380- 1000 nm
EPO-TEK – 377	57	150-300 @100 rpm	≥ 95	0.06 %	102.3	Thermal	Amber/Dark Red	Non-patternable	≥ 90% @ 600nm – 1000 nm
SU-8	52	2005: 6 2015:1500 2100: > 15000	50 (not Cross linked) 210 (Cross linked)	7.5 % @95°C	best	Thermal	Clear	Patternable	> 90% @ 400nm – 800nm

Table 4. 1: Key properties of the bonding materials

4.2.2. Bonding plastic filters to silicon photodetector

During the integration process, it was observed that the optical filters were reacting with SU-8 and other solvents which made them incompatible with the fabrication process. In order to, make them compatible with the integration steps we decided to conformally coat them with a process compatible polymer. Parylene was selected due to its good thermal endurance, zero outgassing and strong resistance to various solvents, acids and alkalis. A 1-1.5 µm thick conformal coating of Parylene-C was obtained by a vacuum coating approach, using PDS 2010, SCS Labcoater. Parylene coated filters were found to be compatible with the fabrication processes. To enhance the bonding strength, the hydrophobic Parylene coated filters were treated with oxygen plasma before bonding. The 120 second oxygen plasma surface activation was carried out in an Oxford Instruments Reactive Ion Etching (RIE) tool (PlasmaLab100). The plasma process used ICP power of 300 W, platen power of 50 W, platen temperature of 15 °C, chamber pressure of 10 mTorr and oxygen mass flow rate of 10 sccm.

After plasma treatment, these filters were immediately subjected to SU-8 bonding as follows. Bonding process begins with cleaning the photodetector dies in acetone and IPA. The photodetector dies were next spin coated with SU-8 2005 at 2000 rpm to obtain a 6-7 μm thick film. Parylene coated filter was bonded manually above the active (n+) region of the photodiode followed by baking the sample on a hot plate at 95 °C for 3 min. During assembly a flat Teflon tweezer was used to squeeze out any trapped air. Such an approach is feasible for flexible substrates only. For brittle substrates this approach of mechanically squeezing out trapped air can lead to device breakage. To bond the second filter over the first one, same process of bonding using SU-8 has been repeated as shown in Figure. 4.1 (b). The alignment accuracy of the first filter to the photo detector active region was ~50 μm and was performed under the microscope. And the filter to filter alignment was ~150 μm. The same can be observed in the SEM image as shown in Figure. 4.5 (d).

4.2.3. Planarization of the bonded filter stack

To enable further stacking of subsequent layers over the filters, the stack needs to be planarized. SU-8 2100 was used for planarizing the filter stack. SU-8 2100 was spin coated over the samples at 3000 rpm for one minute followed by a soft bake at 60 °C for 10 min. The temperature was then ramped to 80° C at a ramp rate of 5 °C min^{-1}. Slow ramping reduces stress issues by ensuring uniform solvent evaporation over the SU-8 thickness. The sample was held at 80 °C for 60 minutes and then slowly cooled to room temperature at a rate of 5 °C min^{-1}. Using a photolithographic mask, the substrates were then exposed to UV light in the EVG 620 mask aligner. The total exposure was broken into 3 steps with each step having a dose of 100 mJ/cm^2. Between each step a gap of 30 seconds was provided to relax and cool the substrate. Such an approach prevents the T-topping[130] of the resist, by reducing the surface heating of the SU-8 film. Post-exposure bake was performed at 60 °C for 10 min and 85 °C for 25 min. The temperature ramp up and ramp were down at a rate of 5 °C min^{-1}. A ten minutes relaxation at room temperature was performed after the soft bake, exposure and post exposure bake steps. Post-bake relaxation helps the photoresist film to attain its optimum hydration levels by reabsorbing the lost moisture. The samples were then developed in SU-8 developer for approximately 22 min with mild agitation. The completion of the development process was judged through optical inspection. Following the IPA/DI water rinse, hard bake was performed at 95 °C for 5 min. Figure. 4.4 (b) shows the photograph after successful bonding of the two-filters on the stack and subsequent planarization.

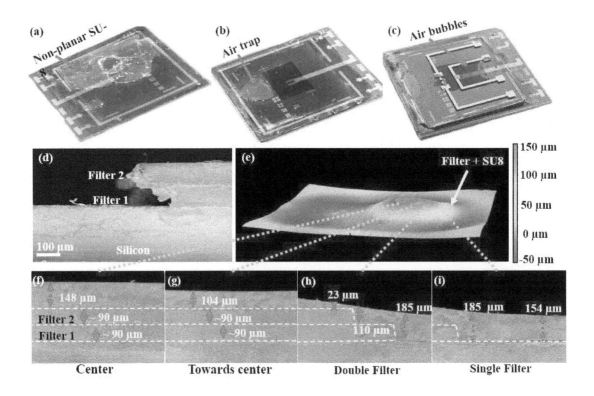

Figure 4. 5: (a) Defects appearing due to non-planar SU-8 after filter stack planarization step. (b) Air traps due to SU-8 outgassing. (c) Air-traps due to SU-8 bonding. (d) SEM image after the filter stacking. (e) 3D mapping of the device after the first planarization step using surface profiler. SEM image at the (f) Centre of the device. (g) Towards the centre of the device. (h) Starting of the double filter. (i) Starting of the single filter.

On some of the sample's planarization failed due to formation of a wavy SU-8 film as shown in Figure. 4.5 (a). This failure was easily detected through optical inspection after the lithography process. Even though the reason for formation of wavy SU-8 film is not very clear, the failure was observed particularly in the stacks where Parylene was damaged at the corners of the filters. If the coated filters were not handled properly, Parylene was observed to peel off from the edges of the filter. This issue was resolved through improving device handling and sorting out the filters with damaged Parylene coating through added inspection steps. Subsequent bonding to such wavy SU-8 layers leads to integration failure from incomplete bonding or appearance of trapped air.

SEM image after bonding the two filters using SU-8 2005 can be seen in Figure. 4.5 (d). After the planarization step 3D topographical map was generated using the Dektak surface profiler. Cross sectional SEM images show the SU-8 thickness at various points after the planarization. SEM image at the centre of the filters can be seen in Figure. 4.5 (f), which shows ~148 μm thick SU-8 above the filter stack. As we move away from the center, SU-8 thickness is reduced to ~ 104 μm as shown in Figure. 4.5 (g). Figure. 4.5 (h) and (i) shows the SEM image at the edge of the double and single filter stacks respectively. From the 3D map and the SEM image it was clear that the SU-8 was higher above the filter and reduces towards the edge of the die, which indicates that SU-8 could not planarize the stack completely. The current planarization step is ensuring a smooth transition of SU-8 from the top of the filters to the photo diode.

4.2.4. Bonding of the glass fluidic chip

To bond the glass wafer to the stack, initially SU-8 was explored. To enhance wettability 1 μm thick Parylene was deposited on the backside of the glass substrate. SU8-2035 was spin-coated above the existing stack consisting of the photodetector and the filters. During the pre-exposure soft bake step while the sample was partially wet, coverslip was manually aligned and bonded under the microscope. However, after the lithography and during the post-exposure/hard bake steps air traps under the glass substrate appeared as shown in Figure. 4.5 (b) and (c). This was attributed to the outgassing of the SU-8 and crosslinking related volume shrinking[131]. SU-8 outgasses at a rate of 7.5% at 95°C (Micro Chem SU-8 2000 data sheet). Most of the stress issues in SU-8 happen due to this crosslinking related volume shrinking. With glass coverslip on the top and the photodetector stack at the bottom, the gas generated gets trapped and leads to bubbles. To address this issue, EPO-TEK epoxies UJ1190 and 377 were considered. EPO-TEK UJ1190 is a UV curable epoxy which has very low outgassing (0.04% at 200° C). A 20 μl drop of epoxy was dispensed over the photodetector stack followed by aligning the glass wafer to the device under the microscope achieving an alignment accuracy in the order of 100 μm using the alignment marks on the photodetector die. The assembled stack was exposed to UV, with a dose of 150 mj/cm² for 5 times at a regular interval of 20 seconds. The exposure was followed by relaxation of 50 min. Figure. 4.4 (d) shows the epoxy bonded sample without any trapped gas. As epoxy wetted the glass surface well, the Parylene coating step was not required. Using EPO-TEK 377 also gave similar results. The remaining non-planarity after the SU-8 step is removed by the epoxy bonding step as seen in Figure. 4.5.

4.2.5. Planarization of the fluidic chip and polymer via opening

Like the first planarization step, SU8-2100 was used again for planarizing the microfluidic device layer. This planarization reduces the step height at the microfluidic device edge and allows formation of electrical routing and interconnects between the photodetector device and the microfluidic device. SU-8 photolithography was performed to open vias to the underlying photodetector layer. The UV exposure energy used in this step was slightly higher, with a dose of 110 mJ/cm² for 3 times and 30 seconds gap between each exposure step. A longer relaxation of 60 min was provided after exposure to ensure minimal thermal stresses in the stack. Finally, the sample was hard baked for 10 min at 110° C. Figure. 4.4 (d) shows the stack after planarization and via opening. Opened vias for metal interconnection between the top microfluidic device and bottom photodetector device can be observed in Figure. 4.6.

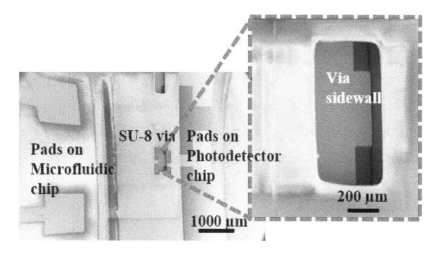

Figure 4. 6: Opened vias after SU-8 lithography

4.2.6. Interconnecting the components in the stack

To electrically connect the top microfluidic chip with the bottom photodetector chip, metallic interconnects are required. Due to the large contour arising from the multiple stacking, bonding and planarization steps, using conventional metal deposition and lithography techniques for forming interconnects is not feasible. Hence, non-conventional techniques like inkjet printing and screen printing were explored. Diamatix 2831 inkjet printer with 10 pL cartridge was used for printing. Silver ink with <50 nm particles was used in this process (Silver dispersion from Sigma Aldrich). Samples were treated with oxygen plasma (300 W – ICP and 50 W – RF, O2-10 sccm, and 10mT) for 60 seconds to enhance adhesion of the silver ink to the glass and SU-8. To increase the thickness of the silver interconnects, the printing was performed 5 times with a gap of 30 min between printing the layers, as shown in Figure. 4.7 (a) and (b). This interval between printing subsequent layers helped the silver ink to get dried as the printer platen was kept at 50 °C. Using the camera on the printer tool the alignment accuracy achieved was in the order of 40 μm in filling the vias.

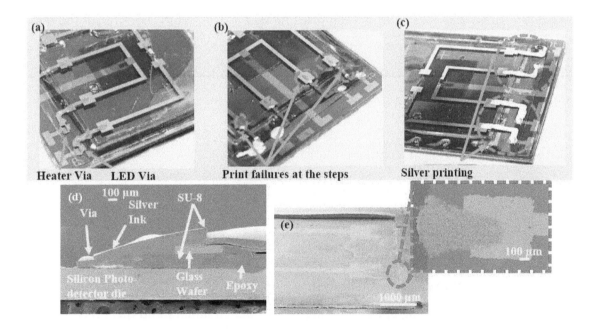

Figure 4. 7: (a) Inkjet printed device (b) Inkjet printed device with defects at the various contours (c) Screen printed device without any defects (d) SEM image of the inkjet printed device showing various contours in the device (e) SEM image after the screen printing.

It was however observed that in some of the samples the connections were not proper. This was due to the high aspect ratio of the SU-8 vias. With a single layer thickness of ~ 440 nm for the inkjet printed silver nano-ink, getting a continuous film on the straight sidewall of the SU-8 via was difficult. Incomplete coverage of silver ink on the SU-8 via sidewall was verified by measuring the local electrical connectivity. Figure. 4.7 (d) SEM image shows the multiple z-height differences in the device. In the cases where the electrical continuity was broken, vias were manually filled with silver epoxy under the microscope as shown in Figure. 4.7 (b).

Due to the limitations of the inkjet process as discussed above, screen printing technique was explored which allows formation of thicker metal lines in a single step. Screen printing allows use of ink with significantly higher nano-particle loading. Such thicker (higher viscosity) inks are incompatible with the inkjet technique. Nylon mesh with the required pattern was purchased from local sources. Silver epoxy was squeezed through the mesh using a hard rubber blade. Micrograph and SEM image of the screen-printed device can be seen in Figure. 4.7 (c) and (e) respectively. The screen-printed epoxy thickness was ~13.9 µm. Screen printing was successful in covering all the bond interfaces as shown in Figure. 4.4 (e) and was found to be repeatable with almost 100% yield in filling the vias. The alignment accuracy was in the order of 150 µm which is slightly poor as compared to the ink jet printing technique.

4.3. Device component testing

4.3.1. Effect of stacking on photodetector sensitivity

Two parylene coated plastic filters were stacked above the photodiode active region using SU-8 2005. This was followed by planarization using SU-8 2100 and stacking of the glass substrate using epoxy bonding technique. To understand the effect of these various materials used in the stacking process spectral response of the device was verified at each stage of the process. For spectral responsivity measurements, a quantum efficiency set-up (QE, Sciencetech) was used. The spectral response of the as fabricated photodetector was measured for wavelengths varying from 400 nm to 700 nm. The sensitivity was found to be ~0.1 A/W as shown in Figure. 4.8 (a). To confirm the functionality of the filters and effect of Parylene on them, spectral responsivity of the photodetector was measured with filters placed on them. Figure. 4.8 (a) shows the measured spectral responsivity of the photodetector at different wavelengths for different configurations. Better blocking of the blue light below 550 nm was observed when a two-filter stack was used instead of one as shown in the measurements (Figure. 4.8 (a)). The slope in the transition region also becomes sharper improving the rejection of the light from the excitation LED. This guided our design to use a two-filter stack. Figure 4. 8(b) shows the spectral response at the various stages of system integration. The measurements were performed on different combination of filter stacks and the final device. From the measurements (Figure. 4.8 (b)) we observe that the 1.5 µm Parylene coating has minimal effect on the optical path. It's clear from the graph that the spectral response of the final device is slightly less than the filters bonded, but the difference is not very high and can be neglected.

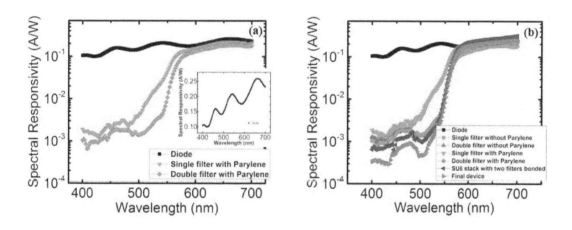

Figure 4. 8: (a) Spectral responsivity measurements of the diode, single and double filter with parylene coating. (b) Spectral responsivity measurements at various stages in the integration process.

4.3.2. Effect of microheater proximity on photodetector sensitivity

The fabricated microheater is in close proximity to the photodetector. To understand the effect of increased temperature on the photodetector sensitivity, spectral responsivity measurements using quantum efficiency set-up (QE, Sciencetech) were performed at various heater voltages as seen in Figure. 4.9. The maximum voltage used translates to a heater temperature of 50 °C (see Figure. 4.10). At lower wavelengths where most of the light is rejected by the filters the increase in noise due to higher temperature is visible in the spectral response. However, at higher wavelengths spectral responsivity results show that the increased temperature has negligible effect on the photodetector response. This is because at higher wavelengths the photodetector current is significantly higher than the noise currents.

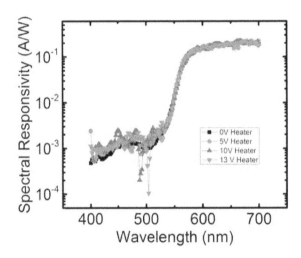

Figure 4. 9: Spectral responsivity measurements of the final device at different heater conditions.

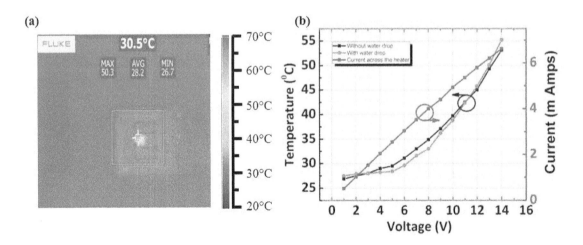

Figure 4. 10: (a) IR image of the device under the test. (b) Temperature Vs Voltage plot of the on chip microheater.

4.3.3. Platform testing

The platinum microheater was calibrated using an IR camera (Fluke). The fabricated die was kept under the camera and the temperature was measured by gradually increasing voltage between the microheater terminals. The microheater voltage vs. substrate temperature is plotted in Figure. 4.10 (b). Measurements of the microheater were performed with and without a 1μL water droplet. Also, the change in current with respect to the applied voltage across the heater is recorded as shown in Figure. 4.10 (b). Based on this graph, it's clear that to obtain heater temperature of ~50 °C a voltage of 14 V has to be applied across the heater. Figure. 4.10 (a) shows the IR camera image of the device.

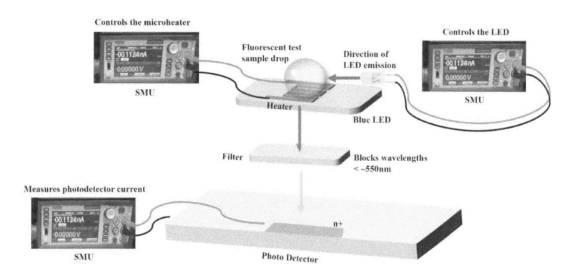

Figure 4. 11: Ray path of the complete setup: the 430 nm peak wavelength blue LED was used as an excitation source and was directed towards the test sample. The light beam falls on the filter stack would allow only the fluoresced light and directs the same towards the photo detector. The change in the current of the photo detector was recorded using the SMU.

Ray path of the complete setup can be seen in Figure. 4.11. Intensity of the blue LED light source was controlled by the source measure unit (SMU) and the light beam from the blue led was used as an excitation source. It is important to note that the integrated LED is facing sideways on the fluidic chip. This makes the emission axis of the LED parallel to the chip surface and hence, allowing illumination of a droplet placed on the fluidic chip. The test sample (Rhodamine) was placed on the top surface of the micro heater. The light is emitted in all directions from the dye droplet. The filters block the wavelengths associated with excitation spectrum while allowing only the fluoresced light from the test sample. The measurements were performed in a closed box as shown in

Figure. 4.12. The fluoresced light will give rise to a change in the current which was recorded using another SMU. The effect on Rhodamine with temperature was also studied by providing voltage to the microheater. The test sample was placed directly on the top surface of the micro heater assembly, to ensure the supplied temperature reach the test sample. After each measurement the surface of the sample was wiped with IPA.

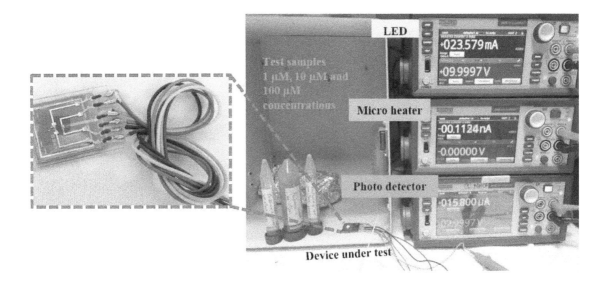

Figure 4. 12: Test setup with 3 source meters for the LED, Microheater and Photodetector, device under test and the Rhodamine dye.

An image of the experimental setup is shown in Figure. 4.12 which consists of wire bonded test sample, three sourcemeters (Kiethley 2460) to supply voltages to photodetector, LED and microheater. Rhodamine B and 6G were obtained as powder from Sigma Aldrich. Rhodamine B has an excitation peak wavelength of λ_{ex}=553 nm and an emission peak wavelength λ_{em}=627 nm. Rhodamine 6G[132] has an excitation peak wavelength of λ_{ex}=526 nm and an emission peak wavelength of λ_{em}=557 nm. 1 µM, 10 µM and 100 µM concentration solutions were prepared in DI water. In our experiments the photodiode measured the intensity of the light emitted by the test samples with and without heating to demonstrate the functionality of the 3D Heterogeneous microfluidic system.

4.4. Experimental Results

A 1 µL drop of Rhodamine 6G dye was placed over the microheater, which is right above the photodetector's active region. Voltage was applied to the LED through a 10 KΩ resistor, this is to ensure there is no high currents to the LED. 3 V to 10 V was applied to the LED using sourcemeter and the photodetector output was recorded at each step for 1 µM, 10 µM and 100 µM concentrations of the solutions. As a control experiment the response was also measured without any dye. The same experiments were repeated for the Rhodamine B fluorescence solution as well. These measurements were performed without applying any voltage to the heater. Figure. 4.13 (a) and (b) shows the results of the Rhodamine 6G and Rhodamine B respectively. These results show that there is a consistent increase in output current of the detector for increasing supply voltage to the LED. Also, with increasing concentration of Rhodamine in both the 6G and B cases, increase in the current has been observed.

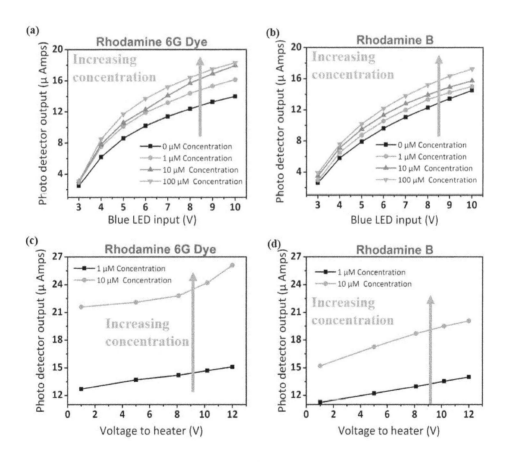

Figure 4. 13: Photodetector output current Vs LED input (a) Rhodamine 6G (b) Rhodamine B. Photodetector output current Vs Voltage to the heater (c) Rhodamine 6G (d) Rhodamine B.

The response of these dyes is sensitive to the solution temperature. Figure. 4.13 (c) and (d) shows the graph of the detector output current for increasing temperatures of Rhodamine 6G and B respectively. In this case the LED voltage was fixed at 10V and the experiments were performed for 1 µM and 10 µM concentration solutions. Results in Figure. 4.13 (c) shows the increase in fluorescence with increasing voltage to the heater for the Rhodamine 6G dye. For the Rhodamine B, contrary to other reports[133,134], we observed an increase in fluorescence with increasing temperature.

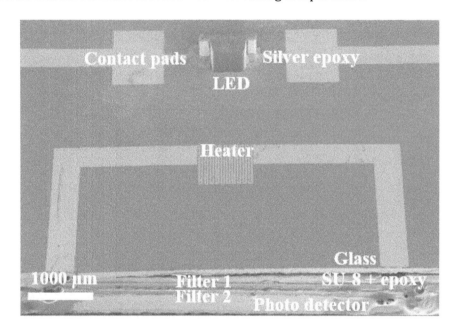

Figure 4. 14: Isometric and the cross-section SEM image of the device with all the components in the system.

The final isometric view SEM image of the device can be seen Figure. 4.14 and the cross section of the stack with the silicon photo detector, both filters, SU8 and epoxy, glass wafer and the final SU-8 which was used to fill the silver epoxy in the vias. Above the existing stack we had bonded another glass cover wafer using epoxy. The stack was planarized again using spin coated thick SU-8 as shown in the false coloured SEM image in Figure. 4.15. This is to show the feasibility of stacking further devices above the existing stack. The unused area on the photodetector die can be used to fabricate the necessary data processing and interface circuitry. Integration of CMOS based detection mechanisms [54,63] and interface circuitry will lead to a significant reduction in the optical and electrical paths. This leads to reduction in both optical and electrical noise. Hence, improving signal to noise ratio. The integration scheme leads to a significant reduction in the volume of the overall system as compared to previous demonstrations[107,113] and hence *achieving scaling of overall system through 3D integration.*

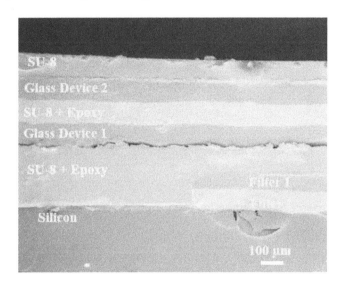

Figure 4. 15: False coloured SEM image showing further stacking of layers

Conclusion

The demonstrated 3D heterogenous integration technology allows packaging of photodetector, filters, microfluidic chip and an LED. This methodology reported here is relatively simple, by allowing integration of commercially available components along with the inhouse fabricated devices, resulting in cost-effective systems with high yield. The key to this approach is planarizing the layers with appropriate SU-8 parameters so that the next layer can be stacked using epoxy without much problem. Implementation of a photo-patternable dielectric (SU-8) as the planarizing layer simplifies formation of vias for achieving interconnects. Using unconventional ink-jet printing and screen-printing of metal nano-inks/epoxies allows the realization of metal interconnects over large topographies. This method paves ways towards possibilities of stacking multiple heterogenous chips on demand. The functionality of the reported heterogenous system was demonstrated by measuring the increase in photo detector current due to the fluorescence property of the Rhodamine B and Rhodamine 6G. Also, the Rhodamine's temperature sensitive property was verified by utilizing the on chip microheater.

Chapter 5
Interconnects

Even though Through Substrate or Silicon Via (TSV) technology is a very well-known technology in the silicon industry, it is still not mature for non-silicon substrates. Using TSV's for 3D integration of heterogeneous dies is currently not feasible. Hence, there is a need to develop new interconnection methodologies using the non-conventional techniques. The main aim of this interconnection approach is to develop a cost-effective interconnect methodology suitable for the heterogenous integration.

In the third and fourth chapters, we have successfully demonstrated 3D vertical stacking approaches for the ultra-thin silicon and GaN based devices. However, these approaches will be more advantages, if these ultra-thin functional device layers were interconnected. In this chapter, we present two approaches for interconnecting stacked layers. The first approach is wire bonding and the second is via filling by ink-jet printing technique. Out of these two approaches ink-jet printing approach is identified to be more promising. This approach allows, stacking of identical sized dies and filling the via holes with 20 µm and higher dimensions which is not possible with the wire bonding approach.

Introduction

Heterogenous Integration of functional platforms like MEMS, microfluidics, optical devices, CMOS, high power GaN/GaAs and RF devices differ in material properties, device dimensions, feature sizes and power dissipation capabilities, is of great importance in the present context of technology. For the interconnects, even though TSV technology is a matured technology in the CMOS industry, it may not be promising for the heterogeneous integration platforms. This is due to the dissimilarities in the material properties and the stresses caused by thermal mismatch and expansion[135]. Another technique to interconnect these layers is by wire bonding approach. Even though it's a popular technique, to enable connections to all the layers, non-identical sized dies only need to be stacked. Non-identical sized dies significantly affect the silicon real estate in the stack[8]. Also, due to the wire bonding pitches, there may be defects in the stacks due to electrical shorting[8].

To address these issues, Ink jet printing by additive manufacturing technique became popular[135-139]. This additive nature of ink-jet printing technique is used for fabrication of micro-meter range conductive tracks and via fills[135,140-142]. Inkjet printing offers significant advantages over the TSV technology in terms of low cost, mask free, ease of alignment and non-contact method of metal deposition[136]. With these game changing advantages, ink jet printing technique became a popular technology for the wearable health care devices[143-147], flexible electronics[148-152], RF devices[137,153-156] and the heterogeneous system integration[145,157-159] platforms.

The previous chapters of this thesis have demonstrated, 3D vertical stacking of ultra-thin silicon stacks and combinations of GaN and silicon stacks. However, these demonstrations lack in interconnection between the dies. Also, non-identical sized dies were used in stacking process, which may affect the silicon real estate. To address these issues, in this chapter we present process flows for interconnection of the ultra-thin silicon dies using the wire bonding and the ink-jet printing techniques.

Here, we use two approaches: (1) Non-identical sized dies were stacked and interconnected using wire bonding technique and (2) identical sized dies were stacked and interconnected using ink-jet printing technique. In both the approaches, functional 10 µm thick device layers from SOI wafer were separated and stacked on to a new carrier wafer. Further, these device layers were interconnected using wire bonding or ink-jet printing techniques. For the ink-jet printing experiments, via fill technique by silver nano particle ink was used to interconnect the layers. Due to the limitations of the standard ink-jet printer tool, it would be difficult to print conductor lines less than 20 µm[140]. To address this issue, process flow with a combination of ink-jet printing for the via fill and metal sputtering for the conductor lines are implemented in this chapter.

5.1. 3D stacking by wafer bonding and interconnection by wire bonding

5.1.1. Fabrication of dummy MEMS dies

5.1.1.1. MEMS silicon cantilever fabrication

An overview of the dummy MEMS device fabrication process flow was shown in Figure. 5.1. From the system integration point of view, a dummy cantilever and two layers of ultra-thin silicon dies were stacked. Piranha cleaned SOI wafer was patterned with the cantilever design and a 5 µm of silicon was etched using the deep reactive ion etching (DRIE) tool from its actual thickness of 10 µm as shown in Figure. 5.1 (a) and (b). This thinning step provides 5 µm gap from the surface of the silicon which helps protecting the cantilever from bonding to the glass wafer during the subsequent bonding steps. Further, the cantilever beam was patterned again and etched using the DRIE tool and was released using the HF vapour phase etcher tool as shown in Figure. 5.1 (c).

5.1.1.2. MEMS glass wafer fabrication for anodic bonding

Amorphous silicon of 1 µm thickness using PECVD was deposited over the piranha cleaned Borofloat glass wafer. Further, Chrome/Gold (Cr/Au) of 20/100 nm thickness was deposited over the amorphous silicon surface as shown in Figure. 5.1 (d). Amorphous silicon and Cr/Au were critical for the HF etching, as they protect the surface of the glass wafer during the subsequent HF wet etching step. It was observed that without amorphous silicon, the glass wafer surface was getting rough during the HF etch. After patterning the glass wafer, Cr/Au and amorphous silicon were etched using KI+I_2 wet etching and the DRIE tool respectively (Figure. 5.1 (e)). Further, using the 50 % Anhydrous HF solution, glass wafer was isotropically etched for 450 seconds to achieve an etch depth of ~55 µm. Finally, leaving the glass surface pristine, Cr/Au and the amorphous silicon were blanket etched (Figure. 5.1 (f)). The purpose of this glass etch, is to protect the cantilever from bonding to glass during anodic bonding.

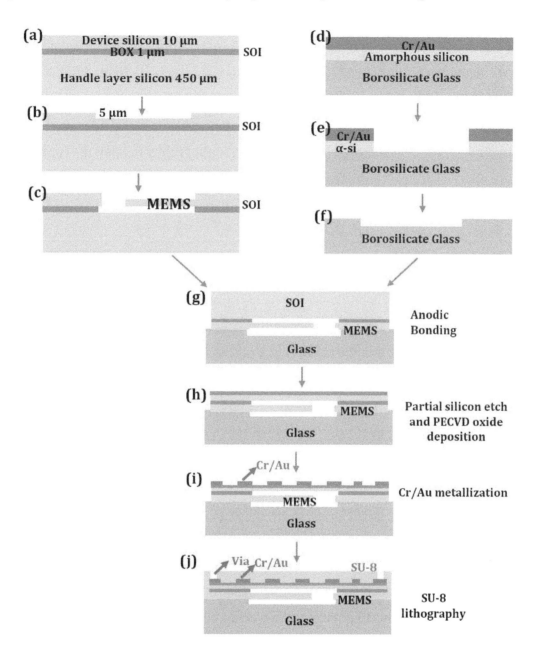

Figure 5. 1: Schematic of the process flow for the dummy MEMS device. (a) SOI wafer used in this process. (b) 5 µm silicon etch using the DRIE tool. (c) Cantilever definition using DRIE tool, followed by cantilever release. (d) Cr/Au and amorphous silicon deposition on the cleaned glass wafer. (e) Patterning and etch of the Cr/Au and amorphous silicon. (f) Wet etch of glass, in the selected region using HF solution. (g) Anodic bonding of the SOI wafer and the glass wafer. (h) Partial thinning of the handle layer silicon to a thickness of 50 µm and 1 µm thick PECVD oxide deposition. (i) 20/100 nm thick Cr/Au metal patterning and etch. (j) SU-8 lithography to open the contact pads on the partially etched handle layer silicon.

5.1.1.3. MEMS cantilever device bonding to the glass

The released SOI wafer was bonded to the Borofloat glass using the anodic bonding technique (Figure. 5.1 (g)). Further, the handle layer silicon was blanket etched using the isotropic silicon etch recipe in the DRIE tool to a silicon thickness of ~ 50 µm from its actual thickness of ~ 450 µm as shown in Figure. 5.1 (h). Using PECVD tool, the surface of the silicon was passivated with 1 µm thick oxide (Figure. 5.1 (h)). Further, lithography and metallization were performed to fabricate the metal lines on the silicon (Figure. 5.1 (i)) which is followed by the SU-8 2015 lithography to open the contact pads as shown in Figure. 5.1 (j). In the interest of system integration point of view dummy MEMS cantilever was chosen. The fabricated cantilever is a dummy device, this doesn't interconnect with the any of the layers.

5.1.2. Fabrication of dummy Ultra-Thin Silicon (UTSi) dies

Here, ultra-thin silicon devices were fabricated by separating the device layer from the SOI wafer. The SOI wafer used in this process was having a device layer of 10 µm, buried oxide of 1 µm and the handle layer of 450 µm thickness. Figure. 5.2 shows the process flow for fabricating UTSi dies. SOI wafer was oxidized using the pyrogenic oxidation technique to obtain a thickness of 1 µm. Cr/Au -20/100 nm thick lines were patterned on the device layer using the lithography and sputtering techniques as shown in Figure. 5.2 (a). Further the SOI wafer was diced to individual die sizes. These dies were temporarily bonded to a carrier silicon wafer using the temporary bonding material Cool Grease CGR 7018 on a hot plate at 65 °C as shown in Figure. 5.2 (b). Using the RIE tool, oxide on the handle layer was etched and using the DRIE tool handle layer silicon was etched completely (Figure. 5.2 (c)). It has been observed that the 1 µm oxide on the device layer is required to balance the stresses with the buried oxide layer, otherwise cracks have been observed in the 10 µm ultra-thin silicon.

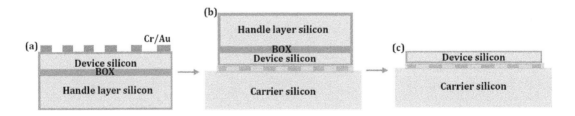

Figure 5. 2: Ultra-thin silicon fabrication. (a) Fabrication of Cr/Au contact lines on the oxidised SOI wafer device layer. (b) Temporary bonding to carrier wafer. (c) Handle layer etch in DRIE tool.

5.1.3. Stacking of the MEMS and silicon devices

Above the SU-8 surface on the MEMS device, the 10 µm ultra-thin silicon device layer was bonded. To enable the wire bonding, the device layers chosen were of different sizes as shown in Figure. 5.3. For the permanent bonding, a 20 µL drop of EPO-TEK 377 epoxy was dispensed on back side of the ultra-thin silicon wafer. The ultra-thin silicon layer along with the carrier was brought in contact with the MEMS die as shown in Figure. 5.3 (b). Bonding was performed for about an hour at 150 °C on a hot plate, and for the basic alignment requirements, optical microscope was utilized. After the bonding, carrier wafer was separated from the stack using the mechanical debonding method followed by thorough cleaning of stack with IPA (Figure. 5.3 (b)). SU-8 2015 lithography was performed to planarize the stack and open up the contact areas on UTSi layer as shown in Figure. 5.3 (c). After the first layer, the same steps were repeated for the 2nd ultra-thin silicon layer bonding and contact area opening using SU-8 as shown in Figure. 5.3 (d) and(e). The final stack can be seen in Figure. 5.3 (e).

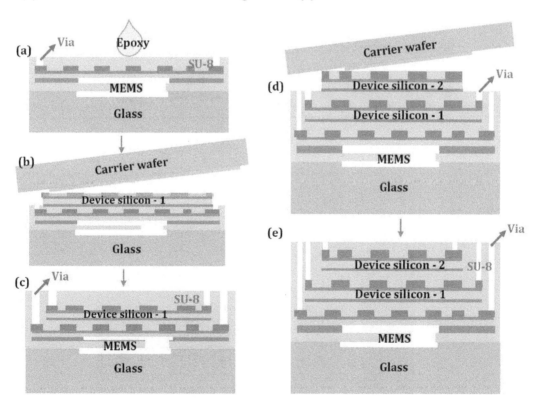

Figure 5. 3: 3D stacking process. (a) Spreading of the epoxy on the SU-8 surface. (b) Permanent bonding of the ultra-thin silicon to the dummy MEMS stack. (c) SU-8 lithography to open the contact regions and planarization. (d) Bonding the second ultra-thin silicon to the existing stack. (e) SU-8 lithography to open the 3rd layer contact pads and planarization.

5.1.4. Stacking results

The SEM image in Figure. 5.4 shows the cross section of the stack including the UTSi layers, inter layer dielectric SU-8 and the epoxy layers. The inset to Figure. 5.4 shows the dummy MEMS cantilever. Figure. 5.5 (a) shows the FIB image of the stack with the two ultra-thin silicon layers. The interconnects for these bonded layers were made to a PCB using wire bonding technique. The photograph of the final PCB with wire bonding can be seen in Figure. 5.5 (b).

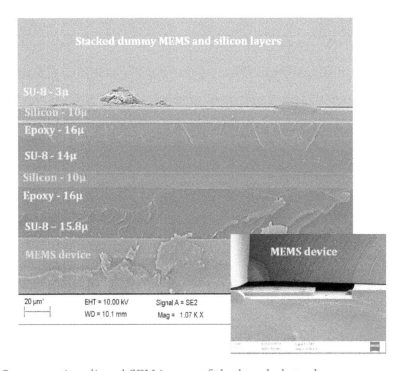

Figure 5. 4: Cross-sectionalional SEM image of the bonded stack.

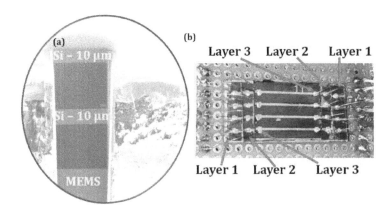

Figure 5. 5: (a) FIB image of the stack. (b) Wire bonded device: metal lines on the dummy MEMS device as layer 1, first ultra-thin silicon layer as layer 2 and the second ultra-thin silicon layer as layer 3.

5.1.5. Temperature cycling of the stacked devices

To check the reliability of the permanent bonding material EPO-TEK 377, the fabricated stack was subjected to thermal cycling in a Votsch temperature cycling chamber. Temperature cycling measurements were recorded for every 1 hour at 25 °C, 0 °C, -40 °C, 120 °C, 85 °C, and 25 °C. The dynamic resistance measurements were recorded for all three layers and were plotted in Figure. 5.6. From the measurements it can be observed that due to higher length of metal lines on layer 1, the resistance was higher than the layer 2 and layer 3. Isolation measurements were performed to check the quality of the inter layer dielectric between the layers and within the same layer. Isolation between the layers and within the same layer is found to be better than 1 GΩ at 100 V DC. Also based on the resistance measurements in Figure. 5.6, the temperature coefficient of resistance is calculated to be ~ 2.5E-3/°C which is observed to be constant for all the layers.

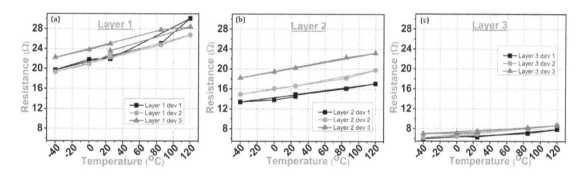

Figure 5. 6: Temperature reliability tests (a) Layer 1: on MEMS die. (b) Layer 2: on first 1st UTSi die. (c) Layer 3: on first 2nd UTSi die.

Further, using the similar process instead of 10 µm device layers, 2 µm device layers were transferred, stacked and interconnected as shown in Figure. 5.7. However, it was observed that during the wire bonding technique, the thin silicon layers were getting cracked as shown in Figure. 5.7. This was found to be due to the bonding force in the wire bonding tool. Hence, for thinner silicon substrates interconnections by wire bonding may not be a feasible idea. Also, using non-identical sized dies is compromising a lot of silicon real-estate and wire bonding multiple interconnects in the stacked dies is found be very challenging.

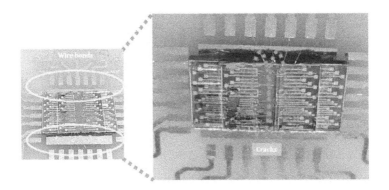

Figure 5. 7: Wire bonding on the UTSi layers with 2 μm thickness

5.2. 3D stacking by Ink jet printed metal via filling

5.2.1. Single layer 3D stacking and interconnection

To address the above discussed challenges, a new process flow has been developed using ink jet printing technique. The Cr/Au metal lines were fabricated on an oxidized SOI wafer before dicing it to 1 cm x 1 cm dies as shown in Figure. 5.8 (a). The SOI wafer utilized in this process has a device layer thickness of 10 μm and buried oxide thickness of 1 μm. Figure. 5.8 and Figure. 5.9 shows the process flow utilized in this transfer process. Individual dies were bonded to glass carrier dies using the temporary bonding material HT:10.10 from Brewer science as shown in Figure. 5.8 (b). The handle layer of the SOI wafer was etched away using the DRIE tool (Figure. 5.8 (c)). On the back side of the etched surface (on buried oxide layer), SU-8 2002 at 5000 rpm was spin coated and baked for 10 min at 120 °C as shown in Figure. 5.8 (a). Further, the surface of the SU-8 was activated with low power oxygen plasma in the RIE tool.

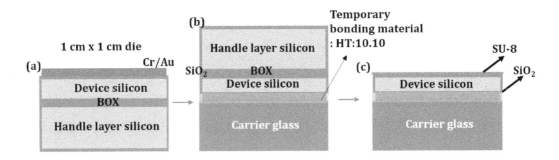

Figure 5. 8: Thin silicon fabrication. (a) SOI wafer with Cr/Au metal lines. (b) Temporary bonding to carrier wafer using HT:10.10. (c) Handle layer silicon etch using the DRIE tool

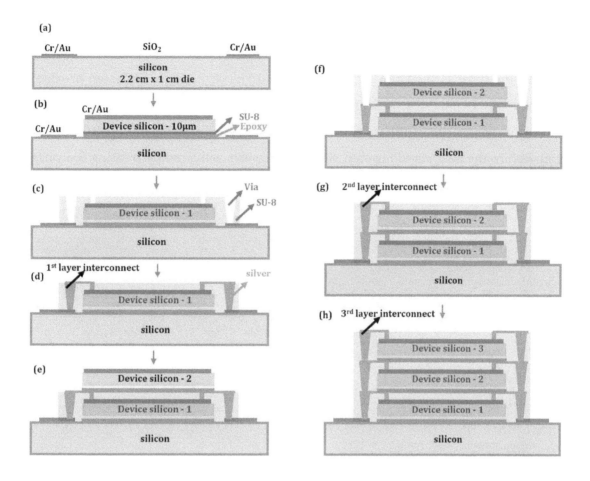

Figure 5. 9: 3D stacking with interconnects process flow. (a) Permanent carrier silicon die with metal pads. (b) Permanent bonding of the ultra-thin silicon to the silicon interposer wafer. (c) SU-8 2005 lithography to open the contact regions and planarization. (d) Via fill using the ink jet printed silver (e) Permanent bonding of the 2nd ultra-thin silicon to the stack. (f) SU-8 lithography to open the 2nd layer contact pads and planarization. (g) Via fill. (h) 3rd layer ultra-thin silicon bonding, SU-8 lithography and the via fill.

A new silicon permanent carrier wafer with Cr/Au contact pads was fabricated and diced to 2.2 cm x 1 cm dies (Figure. 5.9 (a)). Further stacking and interconnections were performed on this die as shown in Figure. 5.9. On the SU-8 surface of the 10 μm device silicon EPO-TEK 377 epoxy was spin coated at 6000 rpm and bonded to the permanent silicon carrier on hotplate at 150 °C for 30 min by bringing both together (Figure. 5.9 (b)). The alignment marks shown in the Figure. 5.10 (a) were used to align the 10 μm die to the permanent carrier die under the optical microscope. After the bonding process, carrier glass wafer was separated from the new stack using the thermal debonding method. This is followed by thorough cleaning of the remaining HT:10.10

material using the wafer bond remover solution. Further, using the SU-8 2005 at 2000 rpm photolithography was performed to planarize the stack and open the contact pads on the new stack as shown in Figure. 5.9 (c) and Figure. 5.10 (a). This step provides a thickness of ~10 µm thick SU-8. The top thin device silicon die has a 50 µm width via and the bottom has 150 µm width via. Both the vias were opened during this SU-8 lithography step. Before the next via filling step, SU-8 surface was activated by treating the surface with low power oxygen plasma.

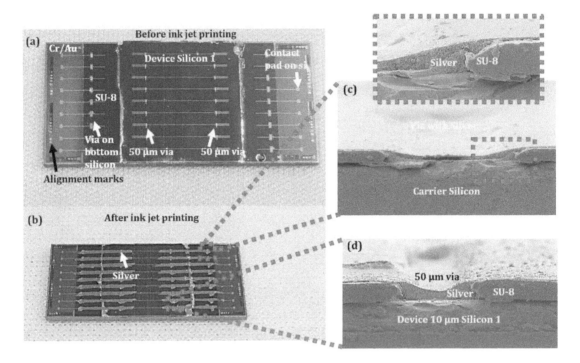

Figure 5. 10: First ultra-thin silicon bonding and interconnects. (a) Ultra-thin silicon bonded to the carrier wafer. (b) Silver ink jet printing the via and the conductor lines. (c) SEM image of the SU-8 via fill on the bottom silicon. (d) SEM image of the SU-8 via fill on the top ultra-thin die with 50 µm via.

5.2.1.1. Ink jet printing of the vias

Ink jet printer from Diamatix was utilized to interconnect the top die and bottom wafer. Vias and the contact lines were printed separately using the printer. Nano particle-based silver ink was used for the interconnections in this process. To fill a 50 µm via with 10 µm thickness, ink jet printing was performed for 5 times. After each printing iteration the device was baked at 150 °C for 10 min to remove the solvents. The number of printing iterations were optimized to partially fill the vias as shown in Figure. 5.10 (c) and (d). Further using the same temperature and time requirements, lines between the vias were printed thrice. Photograph of the printed and interconnected vias can be seen in Figure. 5.10 (b). SEM of the via on the bottom die with inset to the picture showing the silver and SU-8 layers can be seen in Figure. 5.10 (c). Furthermore, 50 µm via fill on the ultra-thin silicon die can be seen in Figure. 5.10 (d).

5.2.2. Multi-layer 3D stacking and interconnection

Using the same process discussed in the section. 5.2.1, another layer of 10 µm thick ultra-thin silicon layer was bonded on to the existing stack as shown in Figure. 5.9 (e). Photograph after the second layer bonding onto the existing stack can be seen in Figure. 5.11 (a). After the bonding step, SU-8 2005 lithography was performed again to open the vias on the bottom silicon and on the top second layer die. Further, again using the same parameters discussed in the previous section, silver ink was filled in the vias and the lines were interconnected as shown in Figure. 5.11 (b). False coloured SEM image in Figure. 5.11 (c) shows the cross section through the 50 µm via after the ink jet printing. The two 10 µm thick silicon layers, inter layer dielectric SU-8 layers, epoxy bonding layer and the ink jet printed silver interconnects for top and bottom die can be seen in the SEM image.

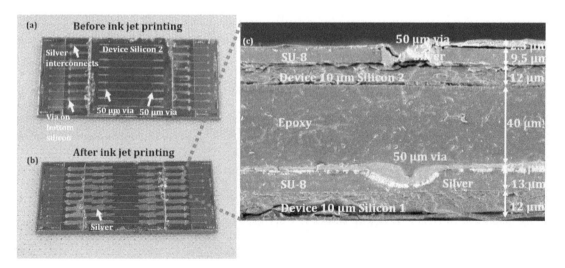

Figure 5. 11: Second ultra-thin silicon bonding and interconnects. (a) second ultra-thin silicon layer bonded to the stack. (b) Ink jet printing of silver in the via and the conductor lines. (c) SEM image of the whole stack showing both the ultra-thin silicon layers and the metal via fills.

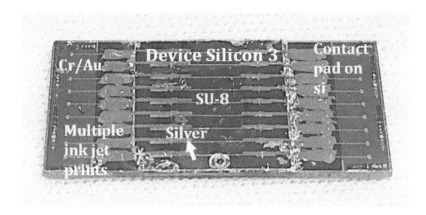

Figure 5. 12: Photograph of the 3-layer stacked device

Further the same stacking process has been repeated with interconnects for the 3-layer stacks as shown in Figure. 5.12. Devices were electrically tested and verified for proper interconnection. However, from Figure. 5.12, it can be observed that the silver ink lines on the bottom die are getting merged. We believe, this is due to the misalignment during the multiple silver ink printing process to be the cause of this. Due to the limitations of the tool, it would be very difficult to control this parameter.

5.2.3. Combination of Ink jet printed Via fill and sputtered metal contacts

It was clear from the previous section that the ink jet printed conductor lines had a higher spread than they were designed to be, which was affecting the pitch of conductor lines. To resolve this issue, we have developed a new process flow with combination of ink jet printing and conventional metal deposition as shown in Figure. 5.13. The metal conductor lines in the device stack were fabricated using the physical vapor deposition technique (sputtering) and vias were filled using the ink-jet printing technique. Metal lines with Cr/Au – 20/100 nm thickness were deposited an oxidized silicon wafer as shown in Figure. 5.13 (a). Further, using the SU-8 2005 lithography contact pads on both ends of the metal lines were opened as shown in Figure. 5.13 (b) and Figure. 5.14 (a). The SU-8 thickness obtained after the lithography was ~ 10 μm. The contact pad opening dimensions were varied from 20 to 50 μm.

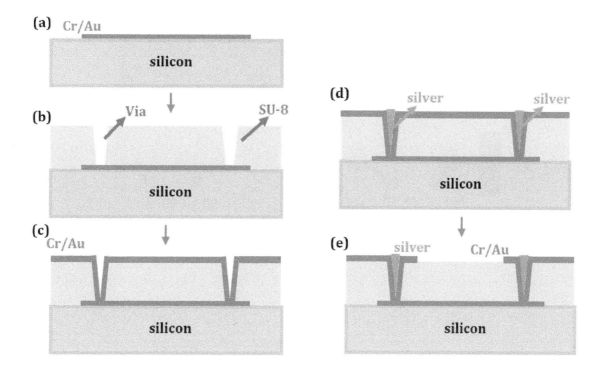

Figure 5. 13: (a) Lift-off of the Cr/Au lines on silicon wafer. (b) 10 μm thick SU-8 lithography and opening of the contact pads. (c) Blanket Cr/Au deposition. (d) Filling the polymer via with silver ink. (e) Lithography and etching of the Cr/Au patterns.

After a brief oxygen plasma on the surface of SU-8, Cr/Au of 20/200nm was deposited by sputtering technique as shown in Figure. 5.13 (c). Further, using the ink jet printing technique the vias were filled 5 times with silver. After each print step, the devices were baked in the oven for 10 min at 150 °C (Figure. 5.13 (d)). Samples were again baked in the oven for an hour after all the via fill steps. Followed by the via fill, photolithography was performed to pattern the Cr/Au region (Figure. 5.14 (b)) and the Cr/Au in the exposed regions was etched using $KI+I_2$ wet etchant solution (Figure. 5.13 (e)). After the Cr/Au etch and the PR strip final device picture can be seen in Figure. 5.14 (c). At this stage the pads were electrically tested and were found to be conducting with a resistance of ~50 Ω. Instead of the Cr/Au the same process has been repeated using the aluminium and it has been observed that Aluminium wet etchant ($H_3PO_4+HNO_3+DI$) is etching the silver and hence devices were getting damaged.

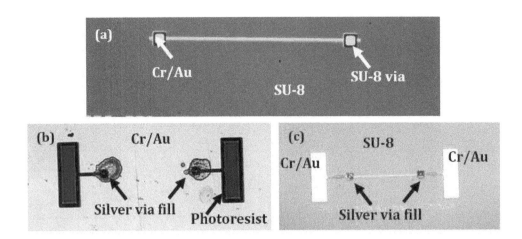

Figure 5. 14: (a) SU-8 lithography. (b) Ink jet via fill and after lithography for Cr/Au etch. (c) Final device after the Cr/Au wet etch and also, bottom Cr/Au is visible through the SU-8.

Based on the results from the above technique, the combination of using inkjet printing for via fill and sputtering for the conductor lines is a viable processing technology for making interconnects in ultra-thin silicon stacks. We are further developing this process for fine pitch interconnections.

Conclusion

In this work, we have demonstrated interconnect processes using the wire bonding and the via fill method. In the wire bonding approach, using the 10 µm thick silicon in the stack gave good results. However, the 2 µm thick device silicon layers were getting cracked during the wire bonding process. Also, it was observed that the interconnections were limited by the wire bond pitches and due to the non-identical size of the dies, effective silicon area was compromised. Whereas, in the ink-jet printing approach, identical sized dies were stacked addressing the effective utilization of silicon real-estate issue. Further, we have demonstrated an interconnect approach involving combination of ink-jet and sputtering techniques which successfully allows for the interconnection of vias with 20 µm and beyond dimensions. Considering the low cost and simplicity involved in the interconnection process, this process could be a great alternative for some of the heterogenous integration applications.

Chapter 5 Interconnects

Chapter 6
Summary and Future Directions

This chapter presents the summary drawn from the results presented in the earlier chapters. It also outlines the scope for further research work.

6.1 Summary

we have developed innovative processing technologies that would allow 3D packaging by the post fab vertical stacking technique, suitable for the packaging industry. We have demonstrated

- In the **second chapter**, we have successfully demonstrated transfer of ultra-thin silicon layers to package substrate (glass) using epoxy and Au-In bonding-based transfer methodologies. Considering the advantages of epoxy bonding over metal bonding, further used epoxy bonding methodology to demonstrate a 3-layer ultra-thin silicon stacking process with functional transistors. The NMOS and PMOS devices on ultra-thin silicon layer have been characterized and compared at various stages of fabrication. Our experimental results exhibit minimal deviation in the post-stacking device behaviour. Based on our simulations and measurements, we concluded the above deviation in current is due to the poor thermal-conduction properties of the foreign substrate. As this stacking process is a post device fabrication process; it can be performed in a backend foundry with no need of changing any front-end foundry processes. Although glass is used as an example of a substrate here, the process outlined here can be replicated on other substrates to facilitate better heat extraction from the devices and/or enable conformal/flexible electronics. This novel simple process may pave way towards 3D IC ultra-thin devices.

- In the **third chapter**, GaN HEMT and MOSFET devices were fabricated and the active device layers were successfully transferred and stacked over a copper substrate. Using a thick electroplated Cu layer, we are able to mitigate the cracking issues arising from the large stress gradients existing in the GaN layer grown on Si substrate. The demonstrated Cu-In approach improved the GaN HEMT device behaviour due the strain induced by the copper electroplating. Very minimal reduction in current has been observed in the case of MOS devices after transfer. This easy and scalable heterogenous stacking of Si-CMOS and III-nitride devices enables 3D Integration of hybrid systems with more functionality, better power, smaller form factor and latency requirements.

- In the **fourth chapter**, the demonstrated 3D heterogenous integration technology allows packaging of photodetector, filters, microfluidic chip and an LED. This reported methodology is relatively simple, by allowing integration of commercially available components, resulting in cost-effective systems with high yield. The key to this approach is planarizing the layers with right SU-8 parameters so that the next layer can be stacked using epoxy without much problem. Using a photo-patternable dielectric (SU-8) as the planarizing layer simplifies formation of vias for formation of interconnects. Using unconventional ink-jet printing and screen-printing of metal nano-inks/epoxies allows formation of metal interconnects over large topographies. This method paves ways towards possibilities of stacking multiple heterogenous chips on demand. The functionality of the reported heterogenous system was demonstrated by measuring the increase in photo detector current due to the fluorescence property of the Rhodamine B and Rhodamine 6G.

- In the **fifth chapter**, we have demonstrated interconnect processes using the wire bonding and the via fill method. In the stacking process, for the wire bonding approach dissimilar sized dies have been utilized and for the via fill method identical sized dies have been used. In the wire bonding approach, using the 10 µm thick silicon in the stack gave excellent results. However, the 2 µm thick device silicon layers were getting cracked during the wire bonding process. The interconnect approach involving a combination of ink-jet and sputtering techniques has given the best results so far. Considering the low cost and simplicity involved in this via fill and interconnection process, this process could be a great alternative for some of the heterogenous integration applications.

6.2 Further scope

- **Heterogeneous Integration of GaN and MOSFET:**
    - This easy and scalable heterogenous stacking of Si-CMOS and III-nitride devices enables 3D Integration of hybrid systems with more functionality, better power, smaller form factor and latency requirements.
    - Interconnection methodologies can be developed to fabricate normally-off GaN HEMT devices. Normally-off silicon devices can be interconnected with normally-on GaN HEMT using the cascode connection. This integration can be performed in the back-end industry.
    - This integration idea can be used in many platforms like high power digital-to-analog converters, wireless transmitters, power switches and silicon control electronics and power amplifiers coupled to silicon differential amplifiers.

- **3D Integration of Heterogeneous Dies for Fluorescent Detection**
    - To get a more efficient device, the silicon photodetector should be replaced with avalanche photodetector or a CMOS based detector, which will enhance the sensitivity of the detection mechanism.
    - The proposed 3D heterogenous integration platform can be used as a fluorescence microscope for detecting specific wavelengths. A detailed study of the possible detectable wavelengths based on the available filters are to be identified.
    - Polymerase chain reaction (PCR) measurements can be performed on this cost-effective setup. A detailed study focusing on PCR measurements for DNA profiling would help significantly to the forensic science area.

- **Interconnects**
    - As mentioned in the last part of the thesis the combination of inkjet for via fill and sputtering for conductor lines, should be a viable idea for making interconnects in ultra-thin silicon stacks.

- As the Heterogenous Integration involves dissimilar materials integration, there is a huge need for better interconnect technologies. Non-conventional ideas like ink-jet printing, screen printing, gravure printing, etc. can be explored.

Bibliography

1. Saraswat, K. Silicon compatible optical interconnects. in *2016 IEEE International Interconnect Technology Conference / Advanced Metallization Conference, IITC/AMC 2016* 1–56 (2016). DOI:10.1109/IITC-AMC.2016.7507739

2. Dennard, R. H. *et al.* Design of Ion-Implanted MOSFETs with Very Small Physical Dimensions. *Proc. IEEE* **87**, 668–678 (1999).

3. Iyer, S. S. & Kirihata, T. Three-Dimensional Integration: A tutorial for designers. *IEEE Solid-state Circuits Magazine*, 63–74 (2015).

4. Knickerbocker *et al.* Three- dimensional silicon integration. *Int. Bus.* **52**, 553–569 (2008).

5. S. S. Iyer, "Heterogeneous Integration for Performance and Scaling," *IEEE Trans. Components, Packag. Manuf. Technol.*, vol. 6, no. 7, pp. 1–10, Feb. 2016, doi: 10.1109/TCPMT.2015.2511626.

6. Moore, G. E. Cramming more components onto integrated circuits (Reprinted from Electronics, pg 114-117, April 19, 1965). *Proc. Ieee* **86**, 82–85 (1965).

7. Beyne, E. The 3-D Interconnect Technology Landscape. *IEEE Des. Test* **33**, 8–20 (2016).

8. Lau, J. H. 3D IC Integration and Packaging. *McGraw Hill Professional,* (2015). ISBN-13: 978-0071848060

9. Christiaens, W., Bosman, E. & Vanfleteren, J. UTCP: A novel polyimide-based ultra-thin chip packaging technology. *IEEE Trans. Components Packag. Technol.* **33**, 754–760 (2010).

10. Iyer, S. S. Three-dimensional integration: An industry perspective. *MRS Bull.* **40**, 225–232 (2015).

11. Fischer, A. C. *et al.* Integrating MEMS and ICs. *Microsystems Nanoeng.* **1**, 15005 (2015).

12. Farooq, M. G. & Iyer, S. S. 3D integration review. *Sci. China Inf. Sci.* **54**, 1012–1025 (2011).

13. Govaerts, J., Bosman, E., Christiaens, W. & Vanfleteren, J. Fine-pitch capabilities of the flat ultra-thin chip packaging (UTCP) technology. *IEEE Trans. Adv. Packag.* **33**, 72–78 (2010).

14. Bleiker, S. J. Heterogeneous 3D Integration and Packaging Technologies for Nano-Electromechanical Systems. PhD thesis *KTH Royal Institute of Technology*, (2017).

15. Ma, Q., Wang, Z. & Pan, L. Monolithic Integration of Multiple Sensors on a Single Silicon Chip. *Symposium on Design, Test, Integration and Packaging of MEMS/MOEMS (DTIP)* 0–3 (2016).

16. Shukri, J.S. 3D ICs Interconnect Performance Modeling and Analysis. PhD thesis

Stanford Unviersity (2012)

17. ITRS. International Technology Roadmap for Semiconductors, Edition 2015, More Moore. 1–52 (2015).

18. Shen, W.-W. & Chen, K.-N. Three-Dimensional Integrated Circuit (3D IC) Key Technology: Through-Silicon Via (TSV). *Nano scale Research letters, (2017)* doi:10.1186/s11671-017-1831-4

19. Motoyoshi, M. Through-Silicon Via (TSV). *Proceedings of the IEEE.* **97**, 43–48 (2009).

20. Ma, M. TSV in IC Packaging : Now and Future. Semicon TAIWAN conference (2003).

21. Eroglu, S. E. K., Cho, W. Y. & Leblebici, Y. A chip-level post-CMOS via-last Cu TSV process for multi-layer homogeneous 3D integration. *2016 12th Conf. Ph.D. Res. Microelectron. Electron. PRIME 2016* 3–6 (2016). doi:10.1109/PRIME.2016.7519491

22. Pangracious, V., Marrakchi, Z. & Mehrez, H. Three-Dimensional Design Methodologies for Tree-based FPGA Architecture. *Springer,* **350**, (2015). ISBN 978-3-319-19174-4

23. Fischer, A. C. *et al.* Integrating MEMS and ICs. *Microsystems Nanoeng.* **1**, 15005 (2015).

24. Knechtel, J., Sinanoglu, O., Elfadel, I. (Abe) M., Lienig, J. & Sze, C. C. N. Large-Scale 3D Chips: Challenges and Solutions for Design Automation, Testing, and Trustworthy Integration. *IPSJ Trans. Syst. LSI Des. Methodol.* **10**, 45–62 (2017).

25. Lau, J. H. Recent Advances and New Trends in Flip Chip Technology. *J. Electron. Packag.* **138**, 030802 (2016).

26. Meyer, T., Ofner, G., Bradl, S., Brunnbauer, M. & Hagen, R. Embedded Wafer Level Ball Grid Array (eWLB). *10th Electron. Packag. Technol. Conf. EPTC 2008* 994–998 (2008). doi:10.1109/EPTC.2008.4763559

27. Lau, J. H. *et al.* Reliability of Fan-Out Wafer-Level Packaging. *IEEE 68th Electronic Components and Technology Conference (ECTC),* 1–12 (2018).

28. R. R. Tummala, Fundamentals of Microsystems Packaging. New York, NY, USA, *Mc Graw-Hill,* 2001, doi: 10.1036/0071371699.

29. Mahajan, R. *et al.* Embedded Multi-die Interconnect Bridge (EMIB)-A High Density, High Bandwidth Packaging Interconnect. *Proc. - Electron. Components Technol. Conf.* 2016–Augus, 557–565 (2016).

30. Wang, C. *et al.* High Bandwidth Application on 2.5D IC Silicon Interposer. *15th International Conference on Electronic Packaging Technology* 568–572 (2014).

31. Hellings, G. *et al.* Active-lite interposer for 2.5 & 3D integration. *IEEE Symp. VLSI Circuits, Dig. Tech. Pap.* 2015–Augus, T222–T223 (2015).

32. Ho, S. W. *et al.* 2.5D through silicon interposer package fabrication by chip-on-wafer (CoW) approach. *Proc. 16th Electron. Packag. Technol. Conf. EPTC 2014* 679–683 (2014). doi:10.1109/EPTC.2014.7028352

33. Wang, M. J. *et al.* TSV technology for 2.5D IC solution. *Proc. - Electron. Components Technol. Conf.* 284–288 (2012). DOI:10.1109/ECTC.2012.6248842

34. Shulaker, M. M. *et al.* Three-dimensional integration of nanotechnologies for computing and data storage on a single chip. *Nature* **547,** 74–78 (2017).

35. Lee, K. H. *et al.* Monolithic integration of Si-CMOS and III-V- on-Si through direct wafer bonding process. **6734,** (2017).

36. Park, J. Physics and Technology of Low Temperature Germanium Mosfets for Monolithic Three Dimentional Integrated Circuits. PhD thesis, *Stanford university* (2009).

37. Deshpande, V. *et al.* Three-dimensional monolithic integration of III – V and Si (Ge) FETs for hybrid CMOS and beyond technology Three-dimensional monolithic integration of III – V and Si (Ge) FETs for hybrid CMOS and beyond. *Jpn. J. Appl. Phys.* **56,** 04CA05 (2017).

38. Ren, J., Member, S., Liu, C. & Tang, C. W. A Novel Si – GaN Monolithic Integration Technology for a High-Voltage Cascoded Diode. *IEEE Electron Device Lett.* **38,** 501–504 (2017).

39. Kerr, A. J., Lee, H. & Palacios, T. Monolithic integration of silicon CMOS and GaN transistors in a current mirror circuit. *Journal of Vacuum Science & Technology B* **101,** (2017).

40. Lee, K. H. K. E. K. *et al.* Monolithic integration of III-V HEMT and Si-CMOS through TSV-less 3D wafer stacking. *Proc. - Electron. Components Technol. Conf.* **2015-July,** 560–565 (2015).

41. E. A. Fitzgerald. Novel Integrated Circuit Platforms Employing Monolithic Silicon CMOS + GaN Devices. *Transactions, E. C. S. & Society.* **75,** 31–37 (2016).

42. Shulaker, M. M. *et al.* Monolithic 3D Integration : A Path From Concept To Reality. *2015 Design, Automation & Test in Europe Conference & Exhibition (DATE)* 1197–1202 (2015). doi:10.7873/DATE.2015.1111

43. Mitsumasa Koyanagi, Kang Wook Lee, Takafumi ukushima andTetsu Tanaka "Challenges in 3D Integration," *ECSTrans.*, vol. 53, 237-244, 2013.

44. Koyanagi, M. Heterogeneous 3D integration - Technology enabler toward future super-chip. *Tech. Dig. - Int. Electron Devices Meet. IEDM* 8–15 (2013). doi:10.1109/IEDM.2013.6724539

45. ITRS. International Technology Roadmap for Semiconductors, Edition 2015, Outside System Connectivity. 1–37 (2015).

46. Garrou, P., Lu, J. J. Q. & Ramm, P. Three-Dimensional Integration. *Handb. Wafer Bond.* 301–328 (2012). doi:10.1002/9783527644223.ch15

47. ITRS. International Technology Roadmap for Semiconductors, Edition 2015, Beyond C-MOS. *Int. Technol. Roadmap Semicond.* (2015).

48. Shulaker, M. M. *et al.* for Computing and Data Storage on a Single Chip. *Nat. Publ. Gr.* **547,** 74–78 (2017).

49. Lee, K. *et al.* A Cavity Chip Interconnection Technology for Thick MEMS Chip

Integration in MEMS-LSI Multichip Module. *J. Microelectromechanical Syst.* **19**, 1284–1291 (2010).

50. ITRS. International Technology Roadmap for Semiconductors 2.0 2015 Edition Heterogeneous Integration. 1–92 (2015). Available: http://www.itrs2.net/itrs-reports.html.

51. Tummala, R., Wolter, K. J., Sundaram, V., Smet, V. & Raj, P. M. New era in automotive electronics, a co-development by Georgia tech and its automotive partners. *2016 Pan Pacific Microelectron. Symp. Pan Pacific 2016* 3–6 (2016). doi:10.1109/PanPacific.2016.7428388

52. ITRS. International Technology Roadmap for Semiconductors, Edition 2015, System Integration. 1–21 (2015).

53. ITRS. International Technology Roadmap for Semiconductors, Edition 2015, Factory Integration. 1–80 (2015).

54. Khan, S. M., Gumus, A., Nassar, J. M. & Hussain, M. M. CMOS Enabled Microfluidic Systems for Healthcare Based Applications. *Adv. Mater.* **30**, 1705759 (2018).

55. ITRS. International Technology Roadmap for Semiconductors, Edition 2015, Executive Report," pp. 79-82, 2015.

56. M. Koyanagi, T. Fukushima, and T. Tanaka, "High-density through silicon vias for 3-D LSIs," *Proc. IEEE*, vol. 97, no. 1, pp. 49–60, Feb. 2009, doi: 10.1109/JPROC.2008.2007463.

57. Zhang, X., Jo, P. K., Zia, M., May, G. S. & Bakir, M. S. Heterogeneous interconnect stitching technology with compressible microinterconnects for dense multi-die integration. *IEEE Electron Device Lett.* **38**, 255–257 (2017).

58. Eroglu, S. E. K., Cho, W. Y. & Leblebici, Y. A chip-level post-CMOS via-last Cu TSV process for multi-layer homogeneous 3D integration. *2016 12th Conf. Ph.D. Res. Microelectron. Electron. PRIME 2016* 3–6 (2016). doi:10.1109/PRIME.2016.7519491

59. Nittala, P. V. K. & Sen, P. Scaling a Fluorescent Detection System by Polymer-Assisted 3-D Integration of Heterogeneous Dies. *J. Microelectromechanical Syst.* PP 1–14 (2018).

60. Temiz, Y., Zervas, M., Guiducci, C. & Leblebici, Y. Die-level TSV fabrication platform for CMOS-MEMS integration. *2011 16th Int. Solid-State Sensors, Actuators Microsystems Conf. TRANSDUCERS'11* 1799–1802 (2011). DOI:10.1109/TRANSDUCERS.2011.5969555

61. J. P. Rojas, G. A. T. Sevilla, M. T. Ghoneim, S. B. Inayat, S. M. Ahmed, A. F. Hussain, and M. M. Hussain, "Transformational Silicon Electronics," *ACS Nano*, vol. 2, pp. 1468–1474, Jan. 2014, doi: 10.1021/nn405475k.

62. Temiz, Y., Lovchik, R. D., Kaigala, G. V & Delamarche, E. Microelectronic Engineering Lab-on-a-chip devices : How to close and plug the lab ? *Microelectron. Eng.* **132**, 156–175 (2015).

63. Nassar, J. M., Sevilla, G. A. T., Velling, S. J., Cordero, M. D. & Hussain, M. M. A CMOS-

compatible large-scale monolithic integration of heterogeneous multi-sensors on flexible silicon for IoT applications. *Tech. Dig. - Int. Electron Devices Meet. IEDM* 18.6.1-18.6.4 (2017). doi:10.1109/IEDM.2016.7838448

64. F. Zimmer, M. Lapisa, T. Bakke, M. Bring, G. Stemme, and F. Niklaus, "One-Megapixel Monocrystalline-Silicon Micromirror Array on CMOS Driving Electronics Manufactured With Very Large-Scale Heterogeneous Integration," *J. Microelectromechanical Syst.*, vol. 20, no. 3, pp. 564–572, Jun. 2011, doi: 10.1109/JMEMS.2011.2127454.

65. Lee, H. S., Ryu, K., Sun, M. & Palacios, T. Wafer-level heterogeneous integration of GaN HEMTs and Si (100) MOSFETs. *IEEE Electron Device Lett.* **33,** 200–202 (2012).

66. Priyabadini, S., Sterken, T., Cauwe, M., Van Hoorebeke, L. & Vanfleteren, J. High-yield fabrication process for 3D-stacked ultrathin chip packages using photo-definable polyimide and symmetry in packages. *IEEE Trans. Components, Packag. Manuf. Technol.* **4,** 158–167 (2014).

67. Kim. S. K., Xue, L. & Tiwari, S. Low-Temperature Polymer-Based Three-Dimensional Silicon Integration. *IEEE Electron Device Lett.* **28,** 706–709 (2007).

68. Ghoneim, M. T. *et al.* Enhanced cooling in mono-crystalline ultra-thin silicon by embedded micro-air channels Enhanced cooling in mono-crystalline ultra-thin silicon by embedded micro-air channels. *AIP Adv.,* **127115,** (2017).

69. Torres Sevilla, G. A. *et al.* High performance high-κ/metal gate complementary metal oxide semiconductor circuit element on flexible silicon. *Appl. Phys. Lett.* **108,** (2016).

70. Priyabadini, S. *et al.* An approach to produce a stack of photo definable polyimide based flat UTCPs. *2012 4th Electron. Syst. Technol. Conf. ESTC 2012* 2–5 (2012). doi:10.1109/ESTC.2012.6542069

71. Krishna, N. P. V. & Sen, P. 3D die level packaging for hybrid systems. in *2016 IEEE International Interconnect Technology Conference / Advanced Metallization Conference, IITC/AMC 2016* (2016). DOI:10.1109/IITC-AMC.2016.7507703

72. Welch, W. C., Chae, J. & Najafi, K. Transfer of metal MEMS packages using a wafer-level solder transfer technique. *IEEE Trans. Adv. Packag.* **28,** 643–649 (2005).

73. Iii, W. C. W. & Najafi, K. Gold-Indium Transient Liquid Phase (Tlp) Wafer Bonding for Mems Vacuum Packaging. *Design. IEEE Trans. Adv. Packag.* 806–809 (2008).

74. Cornelius, W. & Iii, W. Vacuum and Hermetic Packaging, PhD thesis *University of Michigan*, (2008).

75. Galchev, T. V, Welch, W. C. & Najafi, K. A new low-temperature high-aspect-ratio MEMS process using plasma activated wafer bonding. *J. Micromechanics Microengineering* **21,** 045020 (2011).

76. Oberhammer, J. & Stemme, G. BCB contact printing for patterned adhesive full-wafer bonded 0-level packages. *J. Microelectromechanical Syst.* **14,** 419–425 (2005).

77. Chen, S., Griffin, P. B. & Plummer, J. D. Negative Differential Resistance Circuit

Design and Memory Applications. *IEEE Trans. Electron Devices* **56,** 634–640 (2009).

78. Hull, A. W. The dynatron: A vacuum tube possessing negative electric resistance. *Proc. Inst. Radio Eng.* **6,** 5–35 (1918).

79. Chung, J. W., Lee, J., Piner, E. L. & Palacios, T. Seamless On-wafer integration of Si(100) MOSFETs and GaN HEMTs. *IEEE Electron Device Lett.* **30,** 155–156 (2009).

80. Deshpande, V. *et al.* Three-dimensional monolithic integration of III – V and Si (Ge) FETs for hybrid CMOS and beyond technology Three-dimensional monolithic integration of III – V and Si (Ge) FETs for hybrid CMOS and beyond. *Jpn. J. Appl. Phys.* **56,** 04CA05 (1-7) (2017).

81. Zhang, R. *et al.* Silicon-on-insulator with hybrid orientations for heterogeneous integration of GaN on Si (100) substrate. *AIP Adv.* **8,** (2018).

82. Li, X. *et al.* Suppression of the backgating effect of enhancement-mode p-GaN HEMTs on 200-mm GaN-on-SOI for Monolithic Integration. *IEEE Electron Device Lett.* **39,** 999–1002 (2018).

83. Lee, K. H. *et al.* Monolithic integration of Si-CMOS and III-V-on-Si through direct wafer bonding process. *IEEE J. Electron Devices Soc.* **6734,** (2017).

84. Lee, K. H., Bao, S., Fitzgerald, E. & Tan, C. S. Integration of III – V materials and Si-CMOS through double layer transfer process. *Jpn. J. Appl. Phys.* **54,** 1–5 (2015).

85. Hiroki, M. *et al.* Suppression of self-heating effect in AlGaN/GaN high electron mobility transistors by substrate-transfer technology using h-BN. *Appl. Phys. Lett.* **105,** 2014–2015 (2014).

86. Wang, J. *et al.* Thin-film GaN Schottky diodes formed by epitaxial lift-off. *Appl. Phys. Lett.* **110,** (2017).

87. McCall, J. G. *et al.* Fabrication and application of flexible, multimodal light-emitting devices for wireless optogenetics. *Nat. Protoc.* **8,** 2413–2428 (2013).

88. Kim, M., Seo, J.-H., Singisetti, U. & Ma, Z. Recent advances in free-standing single crystalline wide band-gap semiconductors and their applications: GaN, SiC, ZnO, β-Ga_2O_3, and diamond. *J. Mater. Chem. C* **5,** 8338–8354 (2017).

89. Wong, W. S., Sands, T., Cheung, N. W., Sand, T. & Cheung, N. W. Damage-free separation of GaN thin films from sapphire substrates. *Appl. Phys. Lett.* **72,** 599–601 (1998).

90. Chun, J. *et al.* Transfer of GaN LEDs from sapphire to flexible substrates by laser lift-off and contact printing. *IEEE Photonics Technol. Lett.* **24,** 2115–2118 (2012).

91. Zhang, H. A. and Y. B. and E. B. and M. B. and T. B. and P. R. C. and M. C. and K. J. C. and N. C. and R. C. and C. D. S. and M. M. D. S. and S. D. and L. D. C. and B. E. an. The 2018 GaN power electronics roadmap. *J. Phys. D. Appl. Phys.* **51,** 163001 (2018).

92. Park, S. H. *et al.* Wide bandgap III-nitride nanomembranes for optoelectronic applications. *Nano Lett.* **14,** 4293–4298 (2014).

93. Zhang, Y., Leung, B. & Han, J. A liftoff process of GaN layers and devices through

nanoporous transformation. *Appl. Phys. Lett.* **100**, (2012).

94. Temp, B. WaferBOND ® HT-10.10. Online: https://www.brewerscience.com/wp-content/uploads/2016/05/WaferBOND-HT-1010-Data-Sheet.pdf

95. Sheet, T. D. Epo-Tek. **01821**, 1–2 (2015). Online: http://www.epotek.com/site/administrator/components/com_products/assets/files/Style_Uploads/377.pdf

96. Herbecq, N. *et al.* 1900V, 1.6mωcm2 AlN/GaN-on-Si power devices realized by local substrate removal. *Appl. Phys. Express* **7**, 6–9 (2014).

97. Chandrasekar, H., Bhat, K. N., Rangarajan, M., Raghavan, S. & Bhat, N. Thickness Dependent Parasitic Channel Formation at AlN/Si Interfaces. *Sci. Rep.* **7**, 1–10 (2017).

98. Association Connecting Electronic Industries. INDUSTRY STANDARD Moisture / Reflow Sensitivity Classification for Nonhermetic Solid State Surface. *Jt. Ind. Stand.* IPC/JEDEC, 22 (2008).

99. Joh, J. Mechanisms for Electrical Degradation of GaN High-Electron Mobility Transistors. *Change* 5–8 (2006). DOI:10.1109/IEDM.2006.346799

100. Chen, S. *et al.* Millimeter-size single-crystal graphene by suppressing evaporative loss of Cu during low pressure chemical vapor deposition. *Adv. Mater.* **25**, 2062–2065 (2013).

101. Chen, Y. C. & Lee, C. C. Indium-copper multilayer composites for fluxless oxidation-free bonding. *Thin Solid Films* **283**, 243–246 (1996).

102. Lo, G. Q. & Tripathy, S. Raman Scattering and PL Studies on AlGaN / GaN HEMT Layers on 200 mm Si (111). *Eng. Technol.* **6**, 1026–1029 (2012).

103. Shulaker, M. M. *et al.* Carbon nanotube circuit integration up to sub-20 nm channel lengths. *ACS Nano* **8**, 3434–3443 (2014).

104. Abgrall, P. & Gué, a-M. M. Lab-on-chip technologies: Making a microfluidic network and coupling it into a complete microsystem - A review. *J. Micromechanics Microengineering* **17**, R15–R49 (2007).

105. Ghoneim, M. T. *et al.* Enhanced cooling in mono-crystalline ultra-thin silicon by embedded micro-air channels. *AIP Adv.* **5**, (2015).

106. Ryu, G. *et al.* Highly sensitive fluorescence detection system for microfluidic lab-on-a-chip. *Lab Chip* **11**, 1664–1670 (2011).

107. Nelson, N. *et al.* Handheld fluorometers for lab-on-a-chip applications. *IEEE Trans. Biomed. Circuits Syst.* **3**, 97–107 (2009).

108. Kettlitz, S. W., Valouch, S., Sittel, W. & Lemmer, U. Flexible planar microfluidic chip employing a light emitting diode and a PIN- photodiode for portable flow cytometers. DOI:10.1039/c1lc20672a

109. Datta-Chaudhuri, T., Araneda, R. C., Abshire, P. & Smela, E. Olfaction on a chip. *Sensors Actuators, B Chem.* **235**, 74–78 (2016).

110. Chung, J., Hwang, H. Y., Chen, Y. & Lee, T. Y. Microfluidic packaging of high-density CMOS electrode array for lab-on-a-chip applications. *Sensors Actuators B Chem.* **254,** 542–550 (2018).

111. Datta-Chaudhuri, T., Smela, E. & Abshire, P. A. System-on-Chip Considerations for Heterogeneous Integration of CMOS and Fluidic Bio-Interfaces. *IEEE Trans. Biomed. Circuits Syst.* **10,** (2016).

112. Temiz, Y., Kilchenmann, S., Leblebici, Y. & Guiducci, C. 3D integration technology for lab-on-a-chip applications. *Electron. Lett.* **47,** S22 (2011).

113. Datta-Chaudhuri, T., Abshire, P. & Smela, E. Packaging commercial CMOS chips for lab on a chip integration. *Lab Chip* **14,** 1753 (2014).

114. Hussain, A. M. & Hussain, M. M. Deterministic Integration of Out-of-Plane Sensor Arrays for Flexible Electronic Applications. *Small* 5141–5145 (2016). DOI:10.1002/smll.201600952

115. Rae, B. R. *et al.* A CMOS time-resolved fluorescence lifetime analysis micro-system. *Sensors* **9,** 9255–9274 (2009).

116. Dandin, M., Abshire, P. & Smela, E. Optical filtering technologies for integrated fluorescence sensors. *Lab Chip* **7,** 955 (2007).

117. Smela, M. D. and P. A. and E. Polymer filters for ultraviolet-excited integrated fluorescence sensing. *J. Micromechanics Microengineering* **22,** 95018 (2012).

118. James, T. D. *et al.* Valve controlled fluorescence detection system for remote sensing applications. *Microfluid. Nanofluidics* **11,** 529–536 (2011).

119. Lim, J., Gruner, P., Konrad, M. & Baret, J.-C. Micro-optical lens array for fluorescence detection in droplet-based microfluidics. *Lab Chip* **13,** 1472-1475 (2013).

120. Myers, F. B. & Lee, L. P. Innovations in optical microfluidic technologies for point-of-care diagnostics. *Lab Chip* **8,** 2015–2031 (2008).

121. Yao, M., Shah, G. & Fang, J. Highly sensitive and miniaturized fluorescence detection system with an autonomous capillary fluid manipulation chip. *Micromachines* **3,** 462–479 (2012).

122. Long, F. *et al.* Highly sensitive and selective optofluidics-based immunosensor for rapid assessment of Bisphenol A leaching risk. *Biosens. Bioelectron.* **55,** 19–25 (2014).

123. Zhang, R. *et al.* The 2018 GaN power electronics roadmap. *Appl. Phys. Lett.* **8,** 999–1002 (2017).

124. Lafleur, J. P., Jnsson, A., Senkbeil, S. & Kutter, J. P. Recent advances in lab-on-a-chip for biosensing applications. *Biosens. Bioelectron.* **76,** 213–233 (2016).

125. Mata, A., Fleischmann, A. J. & Roy, S. Fabrication of multi-layer {SU-8} microstructures. *J. Micromech. Microeng.* **16,** 276\,-\,284 (2006).

126. Armani, M. *et al.* 2D-PCR: a method of mapping DNA in tissue sections. *Lab Chip* **9,** 3526–3534 (2009).

127. Norian, H., Field, R. M., Kymissis, I. & Shepard, K. L. An integrated CMOS quantitative-polymerase-chain-reaction lab-on-chip for point-of-care diagnostics. *Lab Chip* **14,** 4076–4084 (2014).

128. S. M. Sze, K. K. Ng, Physics of Semiconductor Devices. Hoboken, NJ, USA: *Wiley* 2006..

129. HAMMOND, P. R. Spectra of the lowest excited singlet states of Rhodamine 6G and Rhodamine B. *IEEE J. Quantum Electron.* **15,** 624–632 (1979).

130. Lee, S. J., Shi, W., Maciel, P. & Cha, S. W. Top-edge profile control for SU-8 structural photoresist. *Proc. 15th Bienn. Univ. Ind. Microelectron. Symp. (Cat. No.03CH37488)* 389–390 (2003). doi:10.1109/UGIM.2003.1225777

131. Del Campo, A. & Greiner, C. SU-8: A photoresist for high-aspect-ratio and 3D submicron lithography. *J. Micromechanics Microengineering* **17,** (2007).

132. Tapia Estavez, M. J., Arbeloa, F. L., Arbeloa, T. L. & Arbeloa, I. L. Absorption and Fluorescence Properties of Rhodamine 6G Adsorbed on Aqueous Suspensions of Wyoming Montmorillonite. *Langmuir* **9,** 3629–3634 (1993).

133. Ali, M. a., Moghaddasi, J. & Ahmed, S. a. Temperature Effects in Rhodamine B Dyes and Improvement in CW Dye Laser Performance. *Laser Chem.* **11,** 31–38 (1991).

134. Ahmed, R. M. & Saif, M. Optical properties of rhodamine b dye doped in transparent polymers for sensor application. *Chinese J. Phys.* **51,** 511–521 (2013).

135. Khorramdel, B. & Mäntysalo, M. Fabrication and electrical characterization of partially metallized vias fabricated by inkjet. *J. Micromechanics Microengineering* **26,** 045017 (2016).

136. Khorramdel, B. & Mäntysalo, M. Inkjet filling of TSVs with silver nanoparticle ink. *Proc. 5th Electron. Syst. Technol. Conf. ESTC 2014* 0–4 (2014). DOI:10.1109/ESTC.2014.6962741

137. Putaala, J. *et al.* Capability Assessment of Inkjet Printing for Reliable RFID Applications. *IEEE Transactions on Device and Materials Reliability,* **4388,** (2016).

138. Khorramdel, B., Laurila, M. M. & Mäntysalo, M. Metallization of high density TSVs using super inkjet technology. *Proc. - Electron. Components Technol. Conf.* **2015-July,** 41–45 (2015).

139. Laurila, M., Soltani, A. & Mäntysalo, M. Inkjet Printed Single Layer High-Density Circuitry for a MEMS Device. *IEEE 65th Electronic Components and Technology Conference (ECTC),* 968–972 (2015).

140. Laurila, M., Khorramdel, B. & Mäntysalo, M. Combination of E-Jet and Inkjet Printing for Additive Fabrication of Multilayer High-Density RDL of Silicon Interposer. *IEEE Transactions on electron devices,* 64, (2017).

141. Khorramdel, B. *et al.* Inkjet printing technology for increasing the I / O density of 3D TSV interposers. *Nat. Publ. Gr.* 1–9 (2017). DOI:10.1038/micronano.2017.2

142. Mashayekhi, M. *et al.* Evaluation of Aerosol , Superfine Inkjet , and Photolithography Printing Techniques for Metallization of Application Specific Printed Electronic Circuits. **63,** 1246–1253 (2016).

143. Sillanpää, H., Halonen, E., Liimatta, T. & Mäntysalo, M. Inkjet Printed Wireless Biosensors on Stretchable Substrate. *International Conference on Electronics Packaging (ICEP)*, 322–325 (2014).

144. Sillanpää, H. *et al.* Integration of inkjet and RF SoC technologies to fabricate wireless physiological monitoring system. *Proceedings of the 5th Electronics System-integration Technology Conference (ESTC)*, (2014).

145. Yang, G *et al.* A Health-IoT Platform Based on the Integration of Intelligent Packaging , *IEEE Transactions on Industrial Informatics.* **10,** 2180–2191 (2014).

146. Yang, G., Xie, L., Matti, M., Chen, J. & Tenhunen, H. Bio-Patch Design and Implementation Based on a Low-Power System-on-Chip and Paper-Based Inkjet Printing Technology. *IEEE Transactions on Information Technology and Biomedicine,* **16,** 1043–1050 (2012).

147. Vuorinen, T., Laurila, M.-M., Mangayil, R., Karp, M. & Mäntysalo, M. High Resolution E-Jet Printed Temperature Sensor on Artificial Skin. in *EMBEC & NBC 2017* 839–842 (Springer Singapore, 2018).

148. Vuorinen, T., Niittynen, J., Kankkunen, T., Kraft, T. M. & Mäntysalo, M. Inkjet-Printed Graphene / PEDOT : PSS Temperature Sensors on a Skin-Conformable Polyurethane Substrate. *Nat. Publ. Gr.* 1–8 (2016). doi:10.1038/srep35289

149. Putaala, J. *et al.* Microelectronics Reliability Reliability of SMD interconnections on flexible low-temperature substrates with inkjet-printed conductors. *Microelectron. Reliab.* **54,** 272–280 (2014).

150. Liimatta, T., Halonen, E., Sillanpää, H., Niittynen, J. & Mäntysalo, M. Inkjet Printing in Manufacturing of Stretchable Interconnects. *IEEE 64th Electronic Components and Technology Conference (ECTC)*, 151–156 (2014).

151. Koskinen, S. Electrical Performance Characterization of an Inkjet-Printed Flexible Circuit in a Mobile Application. *IEEE Transactions on Components, Packaging and Manufacturing Technology*, **3,** 1604–1610 (2013).

152. Halonen, E. *et al.* Dynamic Bending Test Analysis of Inkjet-Printed Conductors on Flexible Substrates. *IEEE 62th Electronic Components and Technology Conference (ECTC)*, 80–85 (2012).

153. Myllymäki, S., Putaala, J., Hannu, J., Kunnari, E. & Mäntysalo, M. RF measurements to pinpoint defects in inkjet-printed , thermally and mechanically stressed coplanar waveguides. *Microelectronics Reliability,* **65,** 142–150 (2016).

154. Pynttäri, V., Halonen, E., Sillanpää, H., Mäntysalo, M. & Mäkinen, R. RF Design for Inkjet Technology : Antenna Geometries and Layer Thickness Optimization. *IEEE Antennas and Wireless Propagation Letters,* **11,** 188–191 (2012).

155. Palukuru, V. K. & Sanoda, K. Inkjet-Printed RF Structures on BST – Polymer Composites : An Application of a Monopole Antenna for 2 . 4 GHz Wireless Local Area Network Operation. *International Journal of Applied Ceramic Technology,* **946,** 940–946 (2011).

156. Mäntysalo, M. & Mansikkamäki, P. An inkjet-deposited antenna for 2 . 4 GHz applications. *International Journal of Applied Ceramic Technology,* **63,** 31–35

(2009).

157. Xie, L. *et al.* Heterogeneous integration of bio-sensing system-on-chip and printed electronics. *IEEE J. Emerg. Sel. Top. Circuits Syst.* **2**, 672–682 (2012).

158. Mäntysalo, M. *et al.* System Integration of Smart Packages Using Printed Electronics. 997–1002, I*EEE 62th Electronic Components and Technology Conference (ECTC),* (2012).

159. Mansikkamäki, P., Mäntysalo, M. & Kivikoski, M. Industry-Academy Research Framework on Electronics Hardware Innovations. *Systemics, Cybernetics and Informatics*, **6**, 82–88

Milton Keynes UK
Ingram Content Group UK Ltd.
UKHW032228021224
452012UK00011B/148